(Co)Designing Hope

Laura Cipriani is an assistant professor of landscape architecture at Delft University of Technology. Her current research addresses climate change issues, starting from the materiality of *water* and *soil*, and adopting (co)design approaches. Over the last decade, she has taught at Venice University IUAV, Politecnico di Milano, National University Singapore, Venice International University, and the University of Padua. She holds bachelor's and master's degrees in Architecture (Hons) from IUAV, a Master's in Design Studies (Hons) from Harvard Design School (2004), and a Ph.D. in Landscape Urbanism from IUAV. In 2008, Laura founded Superlandscape, a landscape and urban design firm.

(Co)Designing Hope

Aqueous Landscapes in Transition

Edited by Laura Cipriani

Designed cover image by Laura Cipriani. Photograph by David Cohen.

First published 2025
by Routledge
605 Third Avenue, New York, NY 10158

and by Routledge
4 Park Square, Milton Park, Abingdon, Oxon, OX14 4RN

Routledge is an imprint of the Taylor & Francis Group, an informa business.

Library of Congress Cataloging-in-Publication Data.
A catalog record for this title has been requested.

ISBN: 9781032411460 (hbk)
ISBN: 9781032411477 (pbk)
ISBN: 9781003356486 (ebk)

DOI: 10.4324/9781003356486

Typeset in Garamond Three
by Julian Adda.

Publisher's Note:
This book has been prepared from camera-ready copy provided by the editor.

Peer-review statement:
Two anonymous reviewers selected by Routledge peer reviewed the book structure and Chapter 1; the editor gratefully acknowledges and appreciates their contributions.

Sponsorship:
This publication was made possible by a contribution from the NHBOS Foundation.

STICHTING N·H·BOS
ter bevordering van de landschapsarchitectuur

Nothing is lost, nothing is created, everything is transformed.

Antoine Lavoisier

Contents

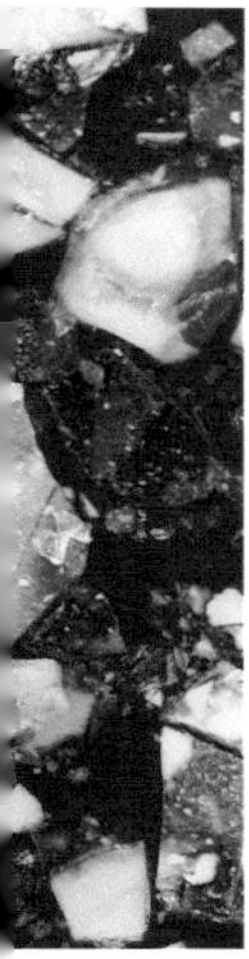

List of Figures

Acknowledgments

This book was initially conceived during the *Crisis and Hope: Landscape as Possibility* conference series organized at TU Delft in the academic year 2020-21. The editorial project started to take shape with the writing of the first chapter. The intent was to put on paper a series of recurring concepts and ideas that emerged about the climate crisis and how landscape design and the collaborative relationships that endure within it can become a hope for future changes. The drafting of the 'axioms' for the landscape in transition was, therefore, the seed of this book, which subsequently led me to contact the contributors for each of the topics outlined.

Firstly, special thanks are due to all the contributors to the book for making this project possible.

Thanks to the editorial staff for organizing the peer review of the book. Thanks to the two anonymous reviewers for their precious feedback on the introductory chapter and the book structure.

Thanks to the professional and dedicated editorial team for their constructive comments and help. Thanks to Kathryn Schell for her editorial support and advice from start to finish on this project. Thanks to Selena Hostetler and Megha Patel for their assistance along the way.

Thanks to Hubert Tonka from Sens&Tonka Editeurs for the reprint permission of Gilles Clément's text and its illustration. Thanks to all the generous photos and image contributors, including Olafur Eliasson, Berndnaut Smilde, and Peter Trautwein for their inspiring projects. Thanks to Julian Adda for the graphic design and his patience in refining the work.

Thanks to Gregory Bracken for organizing the revision of the English text and Peter Kowalski for his incredible help in the editing process.

Many thanks to the generous Stichting NHBOS Foundation for sponsoring the book's graphic design. This work has a different intensity thanks to their financial contribution.

Thank you all.

Contributors

Nadir Abdessemed is a mechanical and aerospace engineer. He was a landscape architecture lecturer for *Scales of Climate* course at Harvard Design School. With Transsolar, he advises integrated design teams for climate-responsive design strategies, from small dwellings to entire masterplans, from clouds at the Venice Biennale to international building projects.

Transsolar
KlimaEngineering
Stuttgart | Germany

Ludwig Berger is a sound artist, composer, and educator with a background in electroacoustic composition at the University of Music Franz Liszt Weimar. Since 2015, he has been a research fellow at the Federal Institute of Technology in Zurich (ETH Zurich), the Swiss Federal Institute of Technology Zurich, where he studies the sonic dimensions of landscapes through experimental field recording.

ETH Zurich
Zurich | Switzerland

Pippa Brashear is the Resilience Principal at SCAPE Landscape Architecture, where she works with multi-disciplinary teams to develop landscape strategies and next-century infrastructure that integrates environmental, economic, and social benefits. Pippa holds a Master's in Landscape Architecture and a Master's in Urban Planning from the Harvard Design School.

SCAPE
Landscape Architecture
New York | NY | USA

Inge Bobbink is an associate professor in landscape architecture at the Faculty of Architecture and the Built Environment at Delft University of Technology (TU Delft). Her current research is about identifying landscape architectonic and sustainable values in traditional water systems worldwide. The goal is to use the acquired knowledge to transform today's water systems into site-specific circular landscapes.

TU Delft
Delft | Netherlands

Amina Chouairi is a Ph.D. student in urbanism at Venice University IUAV. Her research interests focus on landscape design contribution in wet environments and the role of water communities in nature restoration processes, with a particular focus on the Venice Lagoon.

Università IUAV di Venezia
Venice | Italy

TU Delft
Delft | Netherlands

Laura Cipriani is an assistant professor of landscape architecture at Delft University of Technology (TU Delft). Her current research addresses climate change issues, starting from the materiality of *water* and *soil*, and adopting (co)design approaches. She holds bachelor's and master's degrees in Architecture (Hons) from Venice University IUAV, a Master's in Design Studies (Hons) from Harvard Design School (2004), and a Ph.D. in Landscape Urbanism from Venice University IUAV.

ÉNSP
Versailles | France

Gilles Clément is a French agronomist, biologist, writer, entomologist, landscape designer, and teacher at the *École Nationale Supérieure de Paysage* (ENSP Versailles). One of the best-known and most influential landscape architects in Europe, he is the theorist of the *planetary garden*, the *moving garden*, and the concept of the *third landscape*. He has published many essays and books.

University
of Massachusetts, Amherst
Amherst | MA | USA

Carey Clouse is an associate professor in architecture and landscape architecture at the University of Massachusetts, Amherst. She holds a Master of Science in Architecture Studies degree in Architecture and Urbanism from the Massachusetts Institute of Technology and a Bachelor of Architecture from the University of Oregon. Her research addresses climate change adaptation interventions initiated by engaged citizens.

Université de Genève
Geneva | Switzerland

Georges Descombes lives and works in Geneva. He studied architecture in Geneva, Zurich, and London. He is a professor in Geneva and has been a visiting professor at Rapperswil University, Berlage Institute, Harvard Design School, Berkeley University, University of Virginia. The main realizations are Park in Lancy, Swiss Path, Bijlmeer Memorial, and the Aire River project.

Atelier Coloco
Montpellier | France

An Argentinian landscape architect and co-founder of Atelier Coloco, **Miguel Georgieff** lives in France. He studies biology, geography, and landscape design while combining scales and disciplines for operative projects. Atelier Coloco worked with Gilles Clément on the 2022 Landscape Biennale curation team in Versailles under the *Precedence of Living* theme.

Rob Holmes is an associate professor and chair of the Undergraduate Landscape Architecture Program at Auburn University. He is also co-founder of the Dredge Research Collaborative, an independent nonprofit organization that seeks to improve sediment systems through design research, building public knowledge, and facilitating transdisciplinary conversation.

Auburn University
Auburn | AL | USA

Emily Knox is an assistant professor of landscape architecture at Auburn University. She is interested in the role that designers might play in large, complex landscapes. Most recently, her work has involved the ranching landscapes of the American West and the meadow and grassland landscapes of the American Southeast.

Auburn University
Auburn | AL | USA

Annaclaudia Martini is an assistant professor in human geography at the University of Bologna. Her research interests focus on the socio-spatial understanding of post-disaster infrastructure such as seawalls and sea defenses, community recovery, post-disaster tourism, and affective geographies.

University of Bologna
Bologna | Italy

Catherine Mosbach graduated from the *École Nationale Supérieure de Paysage* in Versailles in 1986 after earning an undergraduate degree in Science, Nature, Life, and Physical Chemistry in Strasbourg University (1983). She co-founded *Pages Paysages* magazine (1987–2002) and earned a Master's in History and Civilizations at the *École des Hautes Études en Sciences Sociales* in Paris (1990). She was a visiting professor at the University of Pennsylvania (2003–2006) and Harvard Design School (2017–2019 and 2023).

Mosbach Paysagistes
Paris | France

Transsolar
KlimaEngineering
Stuttgart | Germany

Christine von Raven is a trained architect from the State Academy of Arts and Design Stuttgart. As an urban activist, she is experienced with multi-layered processes and stakeholders. Her focus as an engineer at Transsolar is on complex sustainability strategies and energy concepts at the urban planning level.

UiT
Tromsø | Norway

Eimear Tynan is an associate professor of landscape architecture at UiT, The Arctic University of Norway. She teaches studio and theory courses relating to Arctic and sub-Arctic landscapes. Her main interests are in Arctic islands, Arctic coastal environments, and the ways in which people have engaged with them over time.

LOLA
Landscape Architects
Rotterdam | Netherlands

Peter Veenstra is one of the founding partners of LOLA Landscape Architects in Rotterdam. He studied landscape architecture at Wageningen University. He combines design research with explorative landscape and garden designs. With LOLA Landscape Architects, he published two monographs: *Lost Landscapes* (2014) and *Landscape Works* with Piet Oudolf and LOLA (2021).

MIT
Cambridge | MA | USA

Lizzie Yarina is a doctoral candidate in the Massachusetts Institute of Technology (MIT) Department of Urban Studies and Planning and a research fellow at the MIT Leventhal Center for Advanced Urbanism. Her dissertation investigates the spatial politics of climate change adaptation in Vietnam's Mekong Delta and is supported by a Fulbright-Hays Award. She holds a joint master's degree in Architecture and City Planning from MIT.

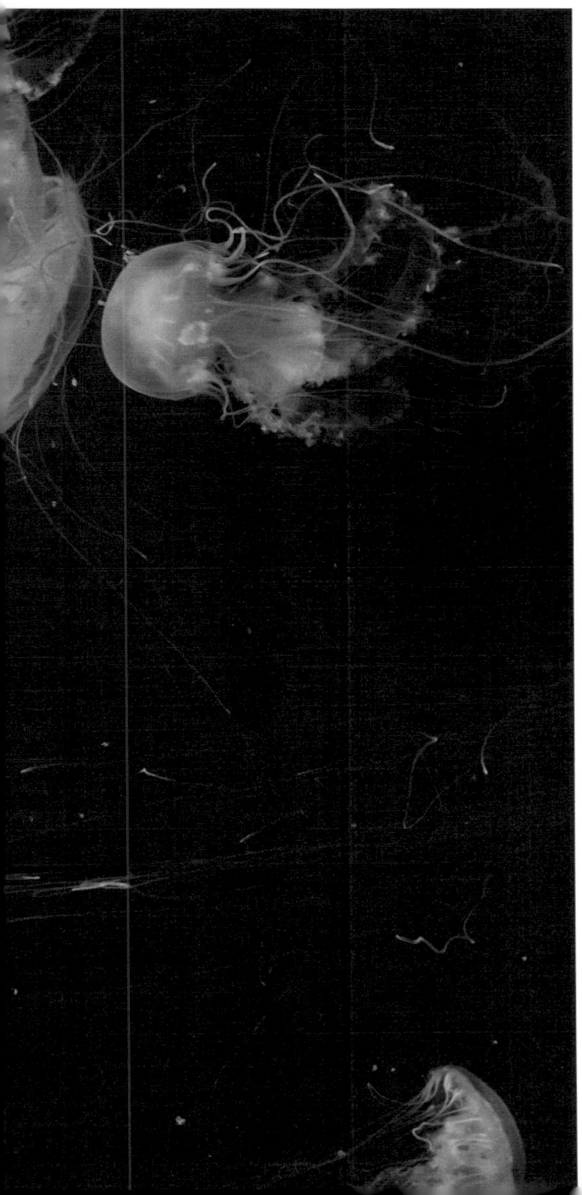

Introduction

1 | Liquid, Solid, and Gas

Axioms for Aqueous Landscapes in Transition

Laura Cipriani

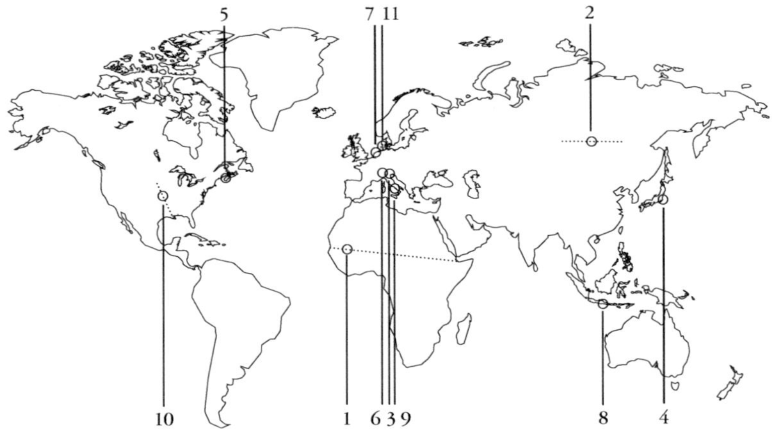

1 | Great Green Wall
Location
From Senegal | Africa
To Gibuti | Africa
Coordinates
From 14° 28' 30.219" N | 14° 27' 10.66" O
To 11° 48' 52.548" N | 42° 50' 43.101" E

2 | Three-North Shelter Forest Program
Location
From Xinjiang | China | Asia
To West Inner Mongolia | China | Asia
Coordinates
From 42° 28' 49.783" N | 85° 27' 48.047" E
To 43° 14' 39.567" N | 114 ° 19' 30.599" E

3 | Woods of San Marco
Location
Cansiglio Forest | Trento | Italy | Europe
Coordinates
46° 3' 48.534" N | 12° 24' 25.815" E

4 | Morino Project
Location
Iwanuma | Prefecture of Miyagi | Japan | Asia
Coordinates
38° 12' 16.768" N | 140° 58' 26.168" E

5 | Living Breakwaters
Location
Tottenville | Staten Island | New York | USA | North America
Coordinates
40° 30' 45.966" N | 74° 15' 7.093" O

6 | Aire River Project
Location
Geneva | Switzerland | Europe
Coordinates
46° 10' 24" N | 6° 05' 03" E

7 | Sand Engine
Location
Monster | The Netherlands | Europe
Coordinates
52° 3' 15.084" N | 4° 11' 22.181" E

8 | EcoShape 'Building with Nature Indonesia'
Location
Demak | Northern Java | Indonesia | Asia
Coordinates
6° 54' 14.097" S | 110° 30' 46.453" E

9 | Dulcis Pompeia Palus
Location
Stabian Plain | Pompeii | Italy
Coordinates
40° 43' 41.179" N | 14° 28' 44.298" E

10 | Mississippi River
Location
From Itasca Lake | Minnesota | USA | North America
To Pilottown, Plaquemines Parish | Louisiana | USA | North America
Coordinates
From 48° 56' 29.292" N | 95° 24' 2.771" O
To 29° 9' 49.445" N | 89° 15' 7.636" O

11 | Wadden Sea
Location
From Den Helder | The Netherlands | Europe
To Mosevrå | Denmark | Europe
Coordinates
From 52° 57' 21.669" N | 4° 43' 39.805" E
To 55° 35' 39.718" N | 8° 15' 12.741" E

Landscape design is the product of a plural, collective symphony that involves or originates from living systems and natural forces. From hydrologist to botanist, fisherman to farmer, swallow to fish, rain to wind, all of Earth's systems constantly draw landscapes. The actors involved in the collaborative dance of the cosmos incessantly design landscapes within a complex and often indecipherable plan.

How can landscape design and different forms of collaboration open new doors to tackle the water and climate crises? What hopes can spring from collective design in its broader meaning?

This introductory chapter sets out a number of 'axioms,' observations, notions, and ideas on landscapes, the water and climate crises, and (co)design, identifying what hopeful routes might be taken for the three states of aqueous landscapes in transition—liquid, solid, and gas.

People, animals, plants, water, ice, fog, wind, sand, and rocks—all contribute to the cosmos' landscape symphony and designing together can become a seed of hope to listen and embrace the Earth's climatic changes.

Crisis and Hope: In Search of a Collaborative Design

We are currently living in a state of crisis identifiable in many parts of the world as economic, environmental, climate-related, political, social, and ethical, with various shades and nuances. We should be prompted by all of these multiple, planetary, and systemic crises to take the necessary and material transformative actions.

Crisis comes from the Greek *krino*, which literally means to *separate*, as in the agricultural operation of separating the wheat from the chaff, the good from the bad, and the seed from the waste. In effect, this is why every crisis brings with it the opportunity for change. As implied by its semantic meaning, a crisis involves a judgment as to what should be kept (as good) and what should be discarded (as detrimental). It represents a turning point, without which the *status quo* would remain unchallenged. Thus, the seed of a crisis is always a source of hope as well.

With various etymological roots, the word *hope* also includes the concept of landscape. The Old English *hōp* indicates a *valley*, especially the upper end of a narrow mountain valley when it is nearly encircled by smooth, green slopes. In Scottish, the term refers to a small bay, an entrance, or a refuge. Landing from the sea or reaching the high parts of mountains induces a feeling of hope in man. The Indo-European root of *hope* (*speh-*) refers to *moving toward a destination* and, therefore, changing direction toward a probable and desirable future—a confident optimism of one's destiny and that of the world. Hope has been linked to the concept of landscape, to an active hope that sets man and natural forces in motion.

Fifty years have passed since *The Limits to Growth* (Meadows et al, 1972), one of the first studies on the various crises that the planet would soon have to face, was published by the Club of Rome. In 1972, a group of scientists from the Massachusetts Institute of Technology launched an appeal to the world after completing studies commissioned by the visionary industrialist Aurelio Peccei that called for a change to the existing model of growth. There is a clear awareness that every crisis—whatever its nature—brings with it a series of multiple planetary crises, somewhat in the manner of a chain reaction. The terms *crisis* and *hope* resound in the text as a warning to society that the world economy should be oriented toward a quest for balance rather than for growth:

> 'Our goal was to provide warnings of potential world *crisis* if these trends are allowed to continue, and thus offer an opportunity to make changes in our political, economic, and social systems to ensure that these *crises* do not take place.'
>
> (Meadows et al, 1972, p. 186)

There is a common denominator to the crises identified, namely the finiteness of natural systems:

> 'The message of this book is urgent and sobering: The Earth's interlocking resources—the global system of nature in which we all live—probably cannot support present rates of economic and population growth much beyond the year 2100, if that long, even with advanced technology. (...) The book contains a message of *hope*, as well: Man can create a society in which he can live indefinitely on Earth if he imposes limits on himself and his production of material goods to achieve a state of global equilibrium with population and production in carefully selected balance.'
>
> (Meadows et al, 1972, p. 210)

Among the many current crises, I see two as being particularly significant. The first, which concerns the climate, is well known and we are now seeing how serious the outlook has become. Extreme events, droughts, floods, shifts in precipitation and temperature patterns, melting glaciers, sea level rise, water salinization, and general changes in the water cycle remind us that the climate crisis is mostly a water crisis. The second—and perhaps even more serious—is a crisis of imagination that is connected with thoughts and creative, far-sighted actions capable of combining the visionary and the pragmatic. Notably, design and designers seem less relevant in the climate discourse. Most books and publications outside the design realm focus on the hazards presented by the climate crisis, rather than on the opportunities it might offer us. Most academic publications prefer to quantify and define the problem, proposing standardized solutions and toolkits rather than offering site-based designs for climate adaptation and mitigation that allow us to (re) think the territory in all its social, economic, and aesthetic dimensions. In this sense, the crisis of imagination relates to the climate crisis.

In my estimation, a response to these two crises can be provided by the discipline of landscape architecture: This has always featured an interdisciplinary approach focusing on water, environmental and ecological issues, the involvement of the community in the design process, and the inclusion of the *time* variable in future plans—yet without neglecting the necessary flexibility of creative and pragmatic thinking.

Taking up these themes, landscape architect and academic Richard Weller published his *Atlas for the End of the World* (Weller, Hoch and Huang, 2017a)—a world map offering a cartographic view of the various planetary-scale crises. Increasing urbanization in the planet's most at-risk bioregions is censused in 38 'hot points of biodiversity,' indicating how future urban expansion across these areas conflicts with natural elements and threatens endangered species. The reporting of systemic crises—water-related, ecological, urban, economic, and social—is presented together with the notion that only wide-ranging and large-scale planning design can effectively *save* the planet. Weller proposed establishing two green wedges around the Earth's circumference, consisting of forests and protected areas: The first on a longitudinal axis extending from North to South America; the second on a transverse axis extending from Europe to Australia via Central and Southeast Asia. Could the landscape indeed become an instrument of change through hope and action? Moreover, could it represent a key factor in gradually overcoming our various planetary crises?

The third term I must refer to is *(co)design* (or *collaborative design*) to highlight landscape design as the product of a plural, collective symphony that involves or originates from living systems and natural forces. From hydrologist to botanist, fisherman to farmer, swallow to fish, rain to wind, all of Earth's systems constantly draw landscapes. The actors involved in the collaborative dance of the cosmos incessantly design landscapes within a complex and often indecipherable plan. Their actions are the conscious products of a will and sometimes the unconscious result of a need, an adaptation, or a chance. Therefore, thinking in (co)design terms means expanding humans' gaze and involves living and non-living things, such as people, animals, plants, water, ice, fog, wind, sand, and rocks, which contribute to the cosmos' landscape symphony. It is a way to feel and catch different sensibilities, starting from different scales: The human and the non-human, the microscopic, and the overall vision.

In the context of collective complex systems, physicist Giorgio Parisi, when studying the flight of flocks (Parisi, 2021), attempted to explain how collective behaviors are derived from the behavior of an individual (e.g., a single bird), which can be interpreted using the laws of physics. Each flock member can communicate to move coherently, producing a single, collective, and plural entity.

> 'At sunset, we see the flocks forming phantasmagorical images; thousands of dancing black spots that stand out against a sky of iridescent colors. We see them all moving together without bumping into each other or dispersing while overcoming obstacles, distancing themselves, and then regrouping, continually reconfiguring their spatial arrangement as if there were an orchestra conductor giving orders that everyone follows. We can spend an indefinite amount of time watching them; the show is always renewed in different and unexpected forms. Sometimes, even in the face of this pure beauty, the professional deformation of a scientist peeps out, and many questions whirl in his head. Is there an orchestra conductor, or is the collective behavior self-organized?'
>
> (Parisi, 2021, p. 7)

The beauty and technique of this natural performance remind us that (co)design requires both of these factors, especially in times of crisis. In the concluding chapter of this book, it is noted that Knox and Holmes (see Chapter 15) referred to the term 'choreography' since the landscape forces can simultaneously design a technical and poetic composition in constructing landscapes.

How can landscape design and different forms of collaboration open new doors to tackle the water and climate crises? The chapters of this book show different scales and levels of design and collaborative water practices: From large governmental projects to small bottom-up interventions; from creative collaboration among designers to traditional community design; from participatory processes to using nature as a (co)designer for tackling the climate crisis. Collaborative design is vital in addressing the climate and water crises, often arising in a state of emergency. Although collective practices do not always result in success, we must remind ourselves that only cooperative and synergic actions can respond to these issues.

Therefore, this book brings together a series of landscape projects concerning water and climate change deriving from the hope of a (co)design shared by the different human and non-human forms involved. The projects represent an attempt to gather different voices worldwide: From researchers to design practitioners. The immediate objectives are to gain a perspective on water landscapes and climate change issues, review notable projects and highlight less-known stories on the topic, focus on the site-specificness of each collaborative practice embedding their cultural aspects, and comprehend how designing together can become a seed for hope. The greater objective is spreading the notion of (co)design being necessary when we discuss climate change while orchestrating narrower and broader changes that attempt to listen to the Earth's possibilities.

Axioms for Acqueous Landscapes in Transition

With acknowledgment of and apologies to Pierce Lewis, author of the well-known article 'Axioms of the Landscape: Some Guides to the American Scene' (Lewis, 1976), I must set out a number of 'axioms,' observations, notions, and ideas on landscapes, the water and climate crises, and (co)design. Defining certain axioms that are concerned with uncertainty and change is no doubt a risky undertaking—oxymoronic, almost—but my feeling is that it helps to crystallize certain familiar issues and identify what routes might be taken when discussing aquatic landscapes in the climate crisis.

1 | *Axiom of (Un)Certainty*

Scientific forecasts of the future climate crisis are typified by uncertainty and inconsistency. Consequently, designing or re-designing the landscape must be redefined in terms of timescales, modes, and methods to account for progressive changeability.

We know that over the course of the coming years and decades, sea levels will rise, extreme weather events will become more frequent and intense,

Figure 1.1 | The collective landscape dance of a flock. Photographs by iStock.com/AGD Beukhof.

Figure 1.2 | A herd of sheep from above. The (co)design is a collective movement. Photograph by iStock.com/Coldsnowstorm.

and various parts of the planet will be affected by temperature changes, the increasing recurrence of rainfall and drought cycles, coastal erosion, and subsidence. While there is a shared awareness regarding climate change currently taking place, we do not know when and how intensely it will be manifested in the coming years. Moreover, the datasets available at present indicate that the dynamics of climate change are evolving faster than initially predicted.

Richard Weller provocatively visualized sea level changes at the global scale if the ice caps were to melt in their entirety. This scenario envisions new imaginary geographies of the world: A sea level rise of 80.32 meters will create different coastlines and inland seas, resulting in the disappearance of more than 50 major cities. 'Even if sea level rises only 0.74 meters by 2100, as is conservatively predicted, some 115 million people will likely be displaced and 420,000 km^2 of land will be lost to the encroaching seas' (Weller, Hoch and Huang, 2017b). We are not aware of what will happen. What we are certain of is that we will face an uncertain future.

2 | *Axiom of Inertia*

Any landscape either remains at rest or in motion unless acted upon by an external force that changes this condition.

The principle of inertia, known also as Newton's first law, can be applied perfectly to the sphere of land and landscape. Every system, of whatever nature, has its own threshold of inertia. As Lewis himself also observed, 'people will not change that landscape unless they are under very heavy pressure to do so' (Lewis, 1976, p. 7).

Locally, but no less globally, no courageous and coordinated *action* is being taken on the part of governments and inhabitants to combat the current threat. Although the consequences of the climate crisis are well known, the causes are only partly understood. Groups of cities exchange ideas by networking and launching initiatives, but there is no concerted implementation of effective and targeted solutions on the territorial scale as yet, whether micro or macro.

Indian writer Amitav Ghosh spoke of a *collective blindness* (Ghosh, 2016). As Ghosh sees it, at the heart of the climate crisis lies a crisis of imagination and cultural awareness. If governments keep referring to shorter and shorter timescales (e.g., five years maximum), then local populations will tend—consciously or unconsciously—to deny or ignore changes that have already been partly evident for some time. For example, this blindness is demonstrated by the fact that properties along coastlines or in areas under threat are still being purchased despite knowledge of future risks. Desirable real estate locations such as Venice and its lagoon continue to be growth markets despite the future climate- and water-related uncertainties.

3 | *Axiom of Catastrophe*

'Most major cultural change does not occur gradually, but instead in great sudden historic leaps—commonly provoked by great events like wars, depressions, major inventions and the like. After these leaps, the landscape is likely to look very different than it did before the leap occurred. Inevitably, however, a lot of "pre-leap" landscape will be left lying around, even though its reason for being has disappeared' (Lewis, 1976, p. 8).

With every rigid system, there comes a moment when a lack of agility and resilience causes the entire mechanism to break down; catastrophic failure, no less. In the 1960s, French mathematician René Thom described natural phenomena by expounding what became known as the 'catastrophe theory' (Thom, 1972; 1983). Thom noted that every element of any given system has a behavior predominantly typified by continuity until a radical, sudden, and abrupt external force causes its condition to change. This evolution is accompanied by earthquakes, volcanoes, instabilities of any nature, and even events of limited magnitude. The catastrophic 'leap' is a sudden passage from one state to another, whereby a system assumes a new and irreversible form. This change occurs over a very short period when compared to the time taken by minimal shifts occurring in more stable states.

Small changes can have huge effects. 'Does the Flap of a Butterfly's Wings in Brazil Set Off a Tornado in Texas?' (Lorenz, 1972) is the title of meteorologist Lorenz's 1972 conference on his theory-poetry of the butterfly effect on how a small change in initial conditions can create a different outcome. Lorenz (1963; 1995) theorized the chaos theory in many natural systems, including weather, climate, and fluid flow.

Water's changes in state provide a good example of this. Let's imagine a glass bottle filled with water. Let's put the bottle in a refrigerator. Water remains a liquid in the bottle over a very long period, while on the surface, everything remains unchanged. Then, at a given moment, the water transforms into ice, the pressure of the ice increases, and the glass bottle breaks open with ice and glass shards bursting suddenly and violently, flowing onto the surface, and creating a new landscape in no time at all—completely changing everything that was there previously.

This point is the catastrophe: An event that is disastrous for the previous state of the system but creates the conditions for the development of a new state in which the system reorganizes itself and continues to evolve until the next catastrophic event occurs. Coastal and river structures show similar behaviors. So-called gray infrastructure can harness rivers within channelized systems or defend coastlines using rigid dikes. When the water volume and pressure are too consistent, any barrier inevitably yields. Rather than fighting against nature, it is always better to adapt to nature, leaving room for water both upstream and along the coast. The famous Dutch project Room for the River (Feddes et al, 2017) is a concrete example of how adaptation must begin in drainage basins by providing adequate space for water in anticipation of extreme events. Thus, the designs of water basins and rivers must include the coast through a series of interventions to mitigate sea erosion, work with coastal profile changes, and sometimes accept an inevitable retreat. Wetlands, dunes, mangrove forests, and coral reefs are the landscape designer's tools, which are used to reduce the intensity of floods and catastrophic events.

4 | *Axiom of Design*

To combat the climate crisis, the design of the landscape is indispensable in order to set in motion a process of action and reaction to occurring phenomena. Every action must be met by an equal and opposite reaction. This 'force' can be represented first and foremost by landscape design and planning. The more that design can be made holistic, the more beneficial the effects will be on the ecological, economic, and social system/ecosystem.

Most of the literature relating to climate change is primarily concerned with investigating and reporting the current problems, postulating possible scenarios, and discussing policies that might be implemented. On the other hand, far less attention has been given to solid planning solutions. Simultaneously, with the main focus on engineering sciences, it seems that the discipline of landscaping is relegated to a role of secondary importance since the 'design,' whatever the scale of intervention, does not appear to constitute the determining variable in planning possible solutions that should not be purely functional but also consider social, environmental, aesthetic, and economic aspects.

In a well-known book by Tomás Maldonado entitled *La Speranza Progettuale* ('The Design Hope') (Maldonado, 1970), the author insisted on the role of design as a tool that is needed to emerge from the ecological and environmental crises that were already becoming evident on a planetary scale during the 1970s. However, he did not advocate comfortable utopian schemes; instead, he suggested taking on the physical, real, and concrete business of design:

> 'It is a choice between destructive pessimism and constructive pessimism: our preference as designers is for the latter. For us, there exists only one option: always and repeatedly to reject everything that can threaten human survival; to help defuse the "time bomb," in other words, to respond to irresponsible growth with responsible control, to respond to congestion with direction. In short: we choose design.'
>
> (Maldonado, 1970, p. 81)

Corollary of Collaboration and Interscalarity

Any plan to combat a crisis—related to climate, especially—can only be implemented through a united effort aimed at establishing a collaborative process whereby institutional and private parties look past the short term and beyond individual interests. The planned interventions must be of interscalar nature: From a single element to regional, national, and global scales.

Large-scale initiatives such as the Great Green Wall in Africa (UNCCD, 2022)—which, as the name suggests, is an imposing barrier (or, in reality, a mosaic) of forests, pastures, and agricultural lands extending from east to west across the Sahara and the Sahel—are designed to operate at the local, regional, and supranational scales. The Great Green Wall program, endorsed by the African Union and involving more than 20 countries, aims to combat growing desertification by mitigating and adapting to the effects of climate change on water cycles. A similar initiative in China is the Three-North Shelter Forest Program, which began in 1978 and aims to plant a long wind-breaking forest that will hold back the expansion of the Gobi Desert (Zhai et al, 2023) while providing the local population with timber.

These large-scale forestry projects have a notable precursor, namely the 15th-century Woods of San Marco in northern Italy, an asset that provided the Venetians with a source of timber and a means of safeguarding the delicate hydrogeological balance of their city's lagoon. The woods of Cansiglio, Somadida, Montello, and Montona in Istria were indeed forestry reserves managed by the Republic of Venice (Moreschi Casti and Zolli, 1988). In the 14th and 15th centuries, with timber being over-harvested by the local populations and with increasingly unstable water levels in the area resulting in more frequent flooding caused by intense deforestation, the Republic was forced to regulate the felling of trees and establish formal cultivation plans to keep the production capacity of the forests on a sound and sustainable footing. Thus, from the 16th century onwards, Venice gave organic structure to a prudent and all-embracing model of governance over its possessions, which included protecting and making the most of woodland and forestry resources.

Even small- or micro-scale projects can contribute to adaptation. Architect Peter Trautwein is building fog nets in several countries around the world (Aqualonis, 2023). Inspired by the traditional experience of fog collectors, the system is based on the condensation of water vapor from the air, transforming fog and dew into drinking water. Similarly, architect Arturo Vittori designed a tower to generate water in arid environments called Warka Tower (Warka Water, 2022). Tested for the first time in 2015 in Ethiopia and later in Cameroon, this device produces 40 to 80 liters of water per day, depending on the weather conditions. Community collaboration is essential since indigenous materials and traditional forms are reinterpreted from time

Figure 1.3 | Making water from fog collectors in Morocco by Peter Trautwein. Photograph by Peter Trautwein. Copyright by Aqualonis.

Figure 1.4 | A detail of the drops of water harvested from fog nets by Peter Trautwein. Photograph by Peter Trautwein. Copyright by Aqualonis.

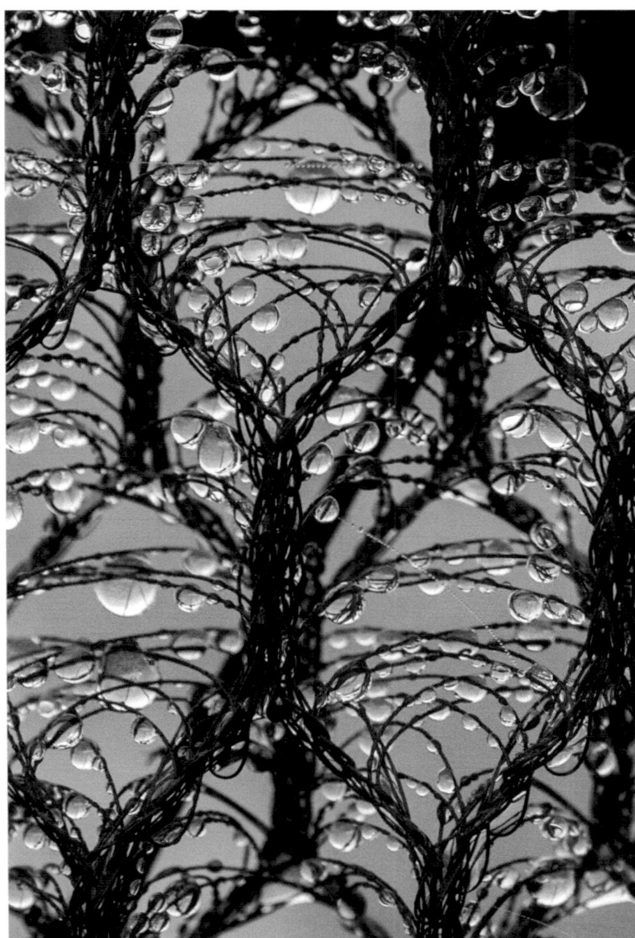

to time in the construction and final design of the Warka Tower. Project participation and management by the people allow for the development of local agriculture and the project's success.

Corollary of (Co)Design by People, Animals, and Nature
To effectively respond to climate urgencies, administrators, institutions, planners, associations, communities, people, living entities, and natural elements must design and (co)design together with the landscape.

Here, I mention three projects that involve collaboration between man and nature to tame the destructive power of water. The first project started with a collaboration among people following a traumatic event. After suffering a devastating earthquake followed by a tsunami on March 11, 2011, in Iwanuma on the coast of Sendai, Miyagi prefecture, Japan, local communities have planted a series of 'forest walls' in the coastal dunes to dissipate the strength of future tidal waves. Taking up the Miyawaki method (Miyawaki, 1999; 2004) and initiating participatory projects sponsored by the Morino Project Foundation (Morino Project, no date), the landscape approach involved the reforestation of coastlines to protect communities. Pioneer species and indigenous trees were planted tightly and haphazardly since the proximity of the plants promotes competition and accelerates the growth process. This method can grow mature forests within 30 years, instead of the 150–200 years it would normally take.

The last two projects started with the (co)design of the landscape architect with natural forces. Sea currents help in the movement of sediments, wind accumulates sand, and animals themselves could contribute to the design of a climate-resilient landscape. Even at the nano-scale level, bacteria, viruses, fungi, algae, plants, and insects can also open the doors to experimentation.

For example, Kate Orff and the office SCAPE developed and experimented with an active oyster reef to protect the New York coastline and purify the water (see Chapter 7). Moreover, the landscape architect Georges Descombes reconfigured the riverbed of the Aire River in Switzerland by exploiting sediment flows (see Chapter 3). The river, which was channelized over centuries, was re-naturalized starting from the design of a creative diamond pattern. The water flowed into the new channels, eroded the riverside, and transported sediments, thereby redefining new riverbed branches. The designer defined a trigger mechanism for the process; in this case, the diamond pattern. Thus, nature, the sediments, and the river are the protagonists of the transformation. Collaboration between man and nature is already a possibility.

Corollary of Design Error
Climate change predominantly affects urban systems that have been developed

in unsuitable areas, such as coastal locations as well as water-endangered and/or geologically hazardous areas. But when town planning choices are made, the elements of the landscape must be given due respect from the initial steps of the design process. Damage repair through remedial landscaping measures will not be sufficient to restore initial conditions.

In his celebrated book *Design with Nature* (McHarg, 1968), architect Ian McHarg highlighted the absolute need for large-scale undertakings in which simple nature-based solutions could counteract environmental imbalances that were already observed. The destruction of natural dune and forest systems in coastal areas and the urbanization that followed was identified by McHarg as the main cause of potential disasters. However, water per se was not the destructive force. Instead, the inadequate interventions of humans provoked sudden and rapid changes. Today, many scientific reports, including those from the Intergovernmental Panel on Climate Change (IPCC, 2021), agree that the climate change we are witnessing is caused by humans. Notably, the Anthropocene era includes the climate and water crises.

Corollary of New Paradigms
The climate crisis prompts and demands a change in the paradigm, design, and tools employed. In short, climate change(s) design.

In recent years, people have begun to see the importance of building *with nature* rather than *against it.* This is a principle discussed by designer Bruno Munari, albeit on a different scale, in his piece *Il Mare Come Artigiano* ('The Sea as a Craftsman'):

> 'You throw something in the sea, and the sea (after an unspecified and indefinable time) returns it to you worked, finished, polished, shiny, or dull depending on the material, and wet too because when wet, the colors are more intense.'
>
> (Munari, 2002, p. 8)

Sea, sand, wind, and nature generally take on the role of craftsmen who shape, mold, and fashion things for humankind. The Sand Engine (De Zandmotor, 2023), an experimental project on the southern coast of the Netherlands, illustrates this principle perfectly: Harnessing natural forces can be of help (e.g., in combating coastal erosion). The project involved creating a substantial artificial peninsula covering an area of approximately 1 km^2, which would be distributed along the entire coastline by exploiting the action of waves, wind, and currents over a period of 20 years. This system will save energy and expense given that in the past, the beaches were artificially replenished with sand every five years as a means of keeping erosion at bay.

Figure 1.5 | The Aire River project by Georges Descombes and Superpositions and its evolution through time from 2016 to 2023. Collage by Laura Cipriani. Images by Esri World Imagery Wayback.

Corollary of Simplicity

The climate crisis is a complex phenomenon that has systemic repercussions on other crises. Design and planning can mitigate complexity through simplicity, starting with simple elements compatible with the peculiarities of the landscape and local populations.

The solutions, which may be similar to others adopted elsewhere, must be devised and adapted to each landscape since every location is unique and non-replicable.

Several projects carried out in Indonesia by the Dutch organization EcoShape show how simple—and sometimes quite basic—devices can bring about an appreciable adaptation of land systems to climate change (EcoShape, 2023). Erosion, subsidence, and rising sea levels are predominant and complex threats to coastal villages in the north of Java. While adopting a strategic vision for the whole area, the work involved placing a number of barriers in the sea, consisting of stakes and brushwood, which were designed to favor the deposition of sediments in order to re-establish a coastal mangrove strip previously destroyed by human activity. Built by the local communities, the elementary structures filter seawater, trapping solids while simultaneously attenuating the force of wave motion. Using this simple, low-tech system, the level of the seabed near the coastline can gradually be raised to allow the natural regrowth of mangroves over time. Reinstatement of the mangrove strip has the effect of defending against coastal erosion, increasing biodiversity by attracting a variety of fish fauna, and purifying the water. Finally, the involvement of local communities in the design and management of the scheme bodes well for its successful long-term implementation.

5 | *Axiom of Time and Timeliness*

The variable of time—short, medium, and long term—is fundamentally important in shaping the design of the future and present. The solutions proposed must advance toward the future through small steps. Like design, the concept of time varies across different cultures. Moreover, design should consider diverse time notions when dealing with water and climate issues.

In the sphere of town planning and landscaping disciplines, the scenario tool has taken on a role of special significance in exploring a future characterized by uncertainty and complexity from a hypothetical standpoint. Planning based on scenarios is a process of answering the question 'What happens if…?,' stimulating a discussion on the future that is reflected indirectly onto the present. Climate-related uncertainties are often addressed through a process of forecasting long-term and very long-term scenarios (2100), whereas the scenario method has rarely been adopted for short- to medium-term studies. For example, if we were to begin planning the future today,

Figure 1.6 | The sand motor project in the Netherlands and its evolution through time from 2012 to 2019. Collage by Laura Cipriani. Images by Esri World Imagery Wayback.

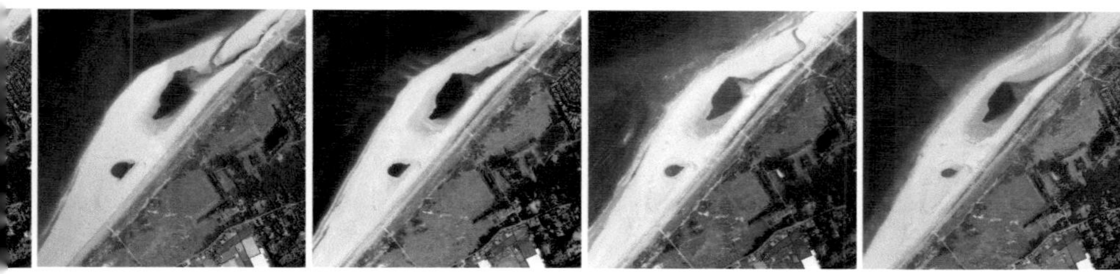

what scenarios would be envisaged for 2025, 2035, and 2050? But more importantly, what will be the scenarios for 2050 if planning is left until an emergency strikes and inevitable disasters have already occurred?

Catastrophe theorist René Thom plainly demonstrated how both *time* and *timeliness* are variables that cannot be ignored if we are not to be dominated by the inevitability of fate:

> '(...) there are many examples of perfectly understood situations in which there is no possible course of action. For example, there is a man on the roof of a flooded house who watches the water rise to inundate him. Inversely, there are situations where one can act efficaciously without understanding why. Where we cannot act, there remains only to show courage against misfortune and to stoically accept destiny's verdict. The world is teeming with situations in which we can clearly intervene, but without really knowing how the effect of our intervention is going to manifest itself.'
>
> (Thom, 1979, p. 11)

Like design, the concept of time varies across different cultures. Design should also consider diverse time notions when we deal with water and climate issues. Western countries mostly rely on a notion of linear time that considers time to be a limited resource. Time is money. Climate change losses urge solutions and a proactive attitude. While South American and Mediterranean countries conceive time as a linear process, the notion implies a multi-active view focusing more on the present rather than on a scheduled process. This might justify a less proactive attitude in planning for future climate actions.

Asian and African countries are mostly based on a cyclical concept that extends to the duration of the day, season, year, and life. This likely implies that climate disasters are part of a cyclical repetition sometimes linked to religious beliefs.

6 | *Axiom of Palimpsests*

Climate events of the past, which have occasionally been catastrophic, are crystallized in palimpsests of the present-day landscape and can provide an important guide to reprogramming the future landscape. History, maps, and the documentation of written sources and documentable iconographic media have a fundamental role to play in the process of designing the future landscape.

A traumatic event such as the eruption of Mount Vesuvius in 79 AD provides a good description of how the landscape can be *read* by studying palimpsests and observing the natural and anthropic traces left over the centuries. Not only did the detritus and mud produced by this disaster destroy and bury the Roman towns of Pompeii, Herculaneum, and Stabiae, but they also changed the geography of the area. The amount of material that erupted was such that the river Sarno changed course toward the east, and the coastline, characterized at the time by a lagoon—the *dulcis Pompeia palus*—facing the river, was transformed into the Stabian Plain of today (Cinque and Russo, 1986).

Similarly, studying the palimpsests of a river system is an essential part of understanding and planning for future climate crises. For example, the celebrated geological mapping of the Mississippi floodplain by Harold Fisk in 1944 (Fisk, 1944) provides an invaluable reconstruction of how the riverbed has evolved over thousands of years. The stages in the life of the watercourse are represented by different colors—green, blue, yellow, red, and orange—in what is almost a work of art as well as science, showing the events and critical changes in the paleo-riverbed.

Therefore, territorial palimpsests are an important aspect of study, research, and design since they can show us the recurring events that took place in the past, presumable changes in the landscape following traumatic events, and possible design solutions for the present and future.

7 | *Axiom of the Cultural Landscape and the Imaginary*

Context matters. The actions required to combat the climate crisis are not universally applicable at all latitudes but will vary depending on the particular situation and cultural landscape. Populations of different cultures respond with different solutions, which are associated with different imaginaries. The imaginary changes not only on a cultural level but also in response to events occurring over time.

Design is always a product of culture, developed in relation to the place where it is rooted. Take the Wadden Sea as an example, which is a mosaic of tidal mud flats along the southeast coastline of the North Sea that creates a long strip of wetland common to three different countries: Netherlands, Germany, and Denmark. These three countries adopt dissimilar coastal defense strategies linked to three distinct cultural approaches since the response of each nation to rising sea levels along their respective seaboards is based on different intentions: Acceptance, adaptation, and resistance. Denmark has abundant coastal areas and no substantial concerns about losing a small amount of land here and there. Accordingly, its policy has been to build a few dikes and allow the sea to shape the future coastline as it may. The Netherlands is a small country with extensive areas of land situated below sea level; therefore, the Dutch are always seeking new methods of adaptation and reclamation: Beaches are replenished with sand, imposing dikes are built, and only where such actions are not strictly feasible will the sea be left to its own devices. Finally, the unbending Germans see the current coastal water line as a front on which they must wage war against the sea, and from which the coast must not recede.

R-I-V-E-R
MISSISSIPPI
ARKANSAS
ASHLEY POINT
HARDEN CUT-OFF
COUNCIL
STATE LAKE

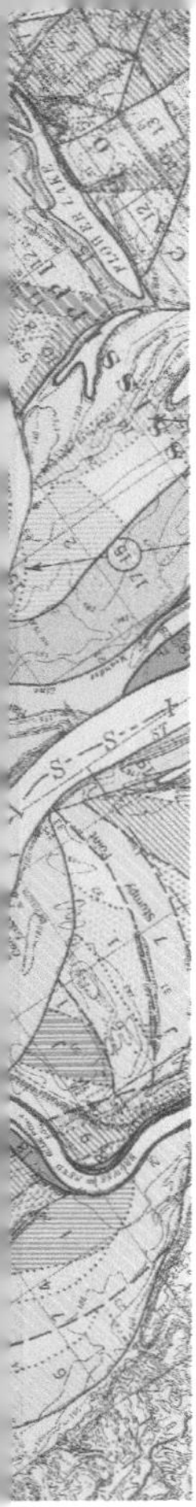

Figure 1.7 | A detail of the colored map of the Mississippi River geological evolution by Harold Fisk. Image by Harold Fisk.

The self-same imagination of the relationship between man and the sea has changed over millennia of history in these parts. The first colonies to be established here approximately 6,000 years ago accepted with equanimity that dwellings, wheat fields, and pastures would often be flooded: The sea was a precious source of nutrients. When the floods became frequent, they raised mounds on which to place their homes, but then the waters began to press hard and they were obliged to build the first earthwork embankments. At this point, the imaginary of the sea began to change: Once a friend, it was now an enemy from which they defended themselves by building a continuous rampart along the coast. And so, the imaginary also changes in response to events.

8 | *Axiom of Climate Injustice and Design Imperialism*

Climate change is a global and a local problem, of which the effects are felt more strongly in certain areas of the planet. Many Asian countries and landscapes and/or developing countries in Africa and South America are those most heavily affected by climate change, despite not being primarily responsible for it.

Consequently, there is a new form of Western imperialism holding sway over many areas of the planet—albeit indirectly and unintentionally—inducing local populations to abandon their homelands and initiating new flows of migration. Based on projected scenarios for population growth and warming over the next 50 years, some authors (Xu et al, 2020) have estimated that between one and three billion people will be living in areas with warmer—and in some cases, extreme—climates. Thus, they will likely migrate toward more temperate zones in a bid to secure the basic necessities of life. Notably, water will become an even scarcer resource.

Paradoxically, in many Asian, Southeast Asian, and other developing regions, we assist in a sort of 'design imperialism' related to water and climate issues. Western projects and knowledge become portable solutions to be applied in other contexts and reveal all their limits: Indifference to local knowledge and wisdom, indifference to local communities' needs, and a climate change design that is not site-based in terms of all its social, economic, and morphological aspects. Design imperialism is a process that is mostly justified by speculation and profit for real estate construction. On this topic, Yarina (see Chapter 5) unveiled how large Western corporations sell inappropriate 'imageries of resilience,' which transform into designs for the climate that are mainly driven by capitalist logic.

9 | *Axiom of Economy*

In plans aimed at mitigating and adapting to the climate crisis, protection from damage must be combined with the production of economic value. A direct economic benefit can provide the incentive needed to overcome the crisis.

It often occurs that damage to the environment caused by climate change is evaluated while ignoring the fact that only a plan capable of generating new economies and innovations can ultimately be effective. Working with less is always preferable. As Gilles Clement stated, working *with* whenever possible and *against* as little as possible is 'a way of thinking that allows (...) to be organized with a view to the greatest economy of means. It presupposes accepting and sometimes even developing ways of collaborating with energies already present, principally those that nature offers in every circumstance and region of the world. It implies being sparing in the consumption of negative energy and if possible, doing without it altogether' (Clement, 2015, p. 34).

10 | *Axiom of Hope*
If catastrophe leads to the failure of one system and the establishment of another, even the simple threat of a catastrophe could bring about positive changes and create new hope.

In an article entitled 'How Climate Change Might Save the World,' Ulrich Beck—risk theorist par excellence—offered a thought that carries hope for the future: The notion that catastrophe becomes a metamorphosis for the whole of society. 'Global risk comes as a threat and brings hope' (Beck, 2014, p. 97). If the proposition that climate change is real becomes definitively accepted, stated Beck:

> 'It induces the necessity (...) to perceive and to practice new forms of transnational responsibility; it puts the problem of cosmopolitan justice on the agenda of international politics; it creates informal and formal cooperation patterns between countries and governments that otherwise ignore each other or even considers themselves enemies: It makes economic and public actors accountable and responsible—even those who do not want to be accountable and responsible. It opens up new world markets, and new innovation patterns, with the consequence being that deniers are losers. It changes lifestyles and consumption patterns; it reveals a strong source of future-oriented meanings in everyday life and for the legitimation of political action (reforms or even revolutions). Finally, it induces new forms of understanding and caring for nature.'
>
> (Beck, 2014, pp. 97–98)

'Active hope' (Macy and Johnstone, 2012) stimulates new practices and a new collective consciousness, as noted by Johanna Macy. People coordinate their actions through a collective thinking process that she describes as 'distributed intelligence' (Macy and Johnstone, 2012, p. 99). Finally, as we will sustain in this book, hope considers the potential of imagination that landscape architecture can drive through a collective (co)designed practice.

Figure 1.8 | Olafur Eliasson and Minik Rosing, *Ice Watch*, 2014. The installation took place at the Bankside, outside Tate Modern in London in 2018. The work raises collective awareness of climate change by providing a direct and tangible experience of the reality of melting arctic ice. Supported by Bloomberg. Photograph by Charlie Forgham-Bailey. Image courtesy of Olafur Eliasson; Neugerriemschneider, Berlin; Tanya Bonakdar Gallery, New York, Los Angeles. Copyright by Olafur Eliasson.

Figure 1.9 | Olafur Eliasson and Minik Rosing, *Ice Watch*, 2014. Listening to the melting ice of *Ice Watch* raises emotional engagement on climate change issues. Supported by Bloomberg. Photograph by Charlie Forgham-Bailey. Image courtesy of Olafur Eliasson; Neugerriemschneider, Berlin; Tanya Bonakdar Gallery, New York, Los Angeles. Copyright by Olafur Eliasson.

Book Structure: Liquid, Solid, and Gas

Today, humans' relationships with water are changing. Water was controlled and regimented in rigid shapes for a long time. Dams and floodgates kept water away in the name of hydraulic safety. However, climate change has prompted a radical change in this design approach. *Working with* rather than *against* water has opened up new possibilities for collaborative design.

This book gathers essays on aquatic landscapes facing climate change challenges and hopes that can spring from collective design in its broader meaning. Through exploring landscapes at different scales, the book focuses on landscape design grounded in different geographies and cultures where (co)design involves all living and non-living systems.

The presented essays deal with aquatic landscapes in transition and are organized into the three states of water as they relate to climate change—liquid, solid, and gas—and their forms. In Latin, the term *form* means *to hold* or *to support*, from which the term *firmus*, which means *stable figure*, is also derived. In the liquid state, water has no shape and assumes a form depending on the support in which it resides. Water is always unstable since it constantly moves in the water cycle. From the soil to the groundwater. From rivers to the ocean. From the ocean to the atmosphere. Evaporation, transpiration, condensation, precipitation, infiltration, surface runoff, and subsurface flow are the physical processes of this constant movement. Today, these changes are becoming increasingly evident. Climate change causes the progressive retreat of glaciers, changes the courses of rivers and sea levels, and upsets water cycles and evaporation phenomena.

When speaking of forms, we refer to the three states of water. However, we can also interpret the term as the Greek *morphé*, which includes perceptual forms of design that arise from the landscape designer's gaze (i.e., the authors of this book).

The opening section, 'Liquid | Water,' discusses collaborative design practices in coastal and riverine areas. Peter Veenstra (see Chapter 2) proposes a collective design manifesto outlining ten climate-adaptive principles drafted with eight landscape design agencies. In his well-known project, Georges Descombes envisions the Aire River as a (co)designer (see Chapter 3). Martini (see Chapter 4) promotes the need to go beyond rigid defense infrastructure in Japan and (co)design for better futures with local communities and natural elements. Addressing Southeast Asian coastal and deltaic megacities, Yarina (see Chapter 5) advocates for overcoming pre-determined models or imported design projects and producing new, local, and collective imaginaries of resilience. Bobbink and Chouairi (see Chapter 6) discuss traditional water systems as the product of collective thinking by local communities. Brashear (see Chapter 7) decomposes the SCAPE manifesto-work 'Living Breakwaters' as a social and ecological engine for collective engagement. Georgieff (see Chapter 8) brings us under the sea to fight the climate crisis through collaboration among researchers, academics, biologists, fishermen, divers, artists, and the silent work of marine living systems.

The second section, 'Solid | Ice,' which might sound like an oxymoron at present, questions the solidity of ice in the era of climate change. The projects noted in this section represent an attempt to raise public awareness and engagement on the issue of climate change while also highlighting unexpected examples of the hopeful adaptation of some local communities.

For example, Tynan (see Chapter 9) proposes a tempo-material approach for (co)designing with time on the Arctic island coast of Bjørnøya. Clouse (see Chapter 10) introduces the climate-adaptive artificial glacier projects of Ladakh, north India, as a way of integrating (co)design by people and natural forces. Berger (see Chapter 11) documents the collective work of researchers and students listening to melting glaciers in the Swiss Alps to raise awareness of our commonality, interconnectedness, and responsibility in the climate crisis.

The final section, 'Gas | Fog and Clouds,' collects design practices related to the third state of water. These include projects and natural phenomena dealing with water and atmospheres, comprising landscape practices and artistic interventions aimed at 'making water' from clouds, fog, and moisture. Gilles Clément and Laura Cipriani (see Chapter 12) describe 'cloud gardeners' that cultivate clouds in dry regions of the planet. Mosbach (see Chapter 13) presents the Phase Shifts Park project to highlight how new water is 'cultivated' by the biomass introduced by it. The water cycle relays life from the depths of roots to the foliage, which evaporates the surplus in the form of oxygen into the sky. Abdessemed and von Raven (see Chapter 14) present their microclimatic installation at the Architectural Biennale in Venice, which features an accessible indoor cloud. This experiment is not far from humans' control of the weather on a global scale.

Finally, the last chapter from Knox and Holmes (see Chapter 15) goes to the roots of the meaning of (co)design, highlighting a symphonic choreography of all planetary forces involved in the water cycle and beyond. An interdependent design dance that can listen to the cosmos' needs and employs aesthetics and techniques to address the climate crisis.

Figure 1.10 | Berndnaut Smilde, *Nimbus Roebourne*, 2017. The artificial clouds of Berndnaut Smilde in Western Australia. Photograph by Bewley Shaylor. Image courtesy of Berndnaut Smilde and Ronchini Gallery.

Figure 1.11 | Berndnaut Smilde, *Nimbus Roebourne*, 2017. The process of making a cloud. Photograph by Annegret Kellner. Image courtesy of Berndnaut Smilde and Ronchini Gallery.

References

Aqualonis (2023) *Floating Springs. Harvesting Drinking Water with Fog Nets.* Available at: <https://www.aqualonis.com/> [Accessed 20 January 2023].

Beck, U. (2014) 'How climate change might save the world.' *Harvard Design Magazine*, 39, pp. 88–98.

Cinque, A. and Russo, F. (1986) 'La linea di costa del 79 d.C. fra Oplonti e Stabiae nel quadro dell'evoluzione olocenica della Piana del Sarno (Campania).' *Bollettino della Società Geologica Italiana*, 105, pp. 111-121.

Clement, G. (2015) *The Planetary Garden: And Other Writings.* Philadelphia, University of Pennsylvania Press.

De Zandmotor (2023) *The Sand Motor. Driver of Innovative Coastal Maintenance.* Rijkswaterstaat, Ministerie van Infrastuctuur en Waterstaat, Provincie Zuid-Holland. Available at: <https://dezandmotor.nl/en/> [Accessed 20 January 2023].

EcoShape (2023) *Building with Nature Indonesia.* Available at: <https://www.ecoshape.org/en/pilots/building-with-nature-indonesia/> [Accessed 20 January 2023].

Feddes, F., Feddes, Y., Luiten, E., Nolden, M. and Sijmons, D. (2017) *Room for the River. Safe and Attractive Landscapes.* Wageningen, Blauwdruk.

Fisk, H.N. (1944) *Geological Investigation of the Alluvial Valley of the Lower Mississippi River.* Vicksburg, MS, Mississippi River Commission, WAR Department Corps of Engineers, U.S. Army. Available at: <https://biotech.law.lsu.edu/climate/mississippi/fisk/fisk.htm> [Accessed 20 January 2023].

Ghosh, A. (2016) *The Great Derangement: Climate Change and the Unthinkable.* Chicago, The University of Chicago Press.

IPCC (2021) 'Summary for policymakers.' In: Masson-Delmotte, V., Zhai, P., Pirani, A., Connors, S.L., Péan, C., Berger, S., Caud, N., Chen, Y., Goldfarb, L., Gomis, M.I., Huang, M., Leitzell, K., Lonnoy, E., Matthews, J.B.R., Maycock, T.K., Waterfield, T., Yelekçi, O., Yu, R. and Zhou, B. eds. *Climate Change 2021: The Physical Science Basis. Contribution of Working Group I to the Sixth Assessment Report of the Intergovernmental Panel on Climate Change.* Available at: <https://www.ipcc.ch/report/ar6/wg1/chapter/summary-for-policymakers/> [Accessed 20 January 2023].

Lewis, P. (1976) 'Axioms of the landscape: Some guides to the American scene.' *Journal of Architectural Education*, 30 (1), pp. 6–9.

Lorenz, E.N. (1963) 'Deterministic nonperiodic flow.' *Journal of the Atmospheric Sciences*, 20, pp. 130–141.

Lorenz, E.N. (1972) *Predictability: Does the Flap of a Butterfly's Wings in Brazil Set off a Tornado in Texas?* Paper presented at the 139th meeting of the American Association for the Advancement of Science, Washington DC. Available at: <http://eaps4.mit.edu/research/Lorenz/Butterfly_1972.pdf> [Accessed 20 January 2023].

Lorenz, E.N. (1995) *The Essence of Chaos.* Seattle, University of Washington Press.

Macy, J. and Johnstone, C. (2012) *Active Hope: How to Face the Mess We're in Without Going Crazy.* Novato, California, New World Library.

Maldonado, T. (1970) *La Speranza Progettuale: Ambiente e Società.* Torino, Einaudi.

McHarg, I.L. (1968) *Design with Nature.* New York, The Natural History Press, Garden City.

Meadows, D.H., Meadows, D.L., Randers, J. and Behrens III, W.W. (1972) *The Limits to Growth: A Report for the Club of Rome's Project on the Predicament of Mankind.* New York, Universe Books.

Miyawaki, A. (1999) 'Creative ecology: restoration of native forests by native trees.' *Plant Biotechnology*, 16 (1), pp. 15–25.

Miyawaki, A. (2004) 'Restoration of living environment based on vegetation ecology: Theory and practice.' *Ecological Research*, 19, pp. 83–90.

Moreschi Casti, E. and Zolli, E. (1988) *Boschi della Serenissima: Storia di un Rapporto Uomo-Ambiente.* Venezia, Arsenale editrice.

Morino Project (no date) *Project Summary.* Available at: <https://morinoproject.com/english> [Accessed 20 January 2023].

Munari, B. (2002) *The Sea as a Craftsman.* Mantova, Corraini Edizioni.

Parisi, G. (2021) *In un Volo di Storni: Le Meraviglie dei Sistemi Complessi.* Milano, Mondadori Libri S.p.A.

Thom, R. (1972) *Stabilité Structurelle et Morphogenèse.* Paris, InterÉditions.

Thom, R. (1979) 'At the boundaries of man's power: Play.' *SubStance*, vol. 8, 4 (25), pp. 11–19.

Thom, R. (1983) *Paraboles et Catastrophes.* Paris, Flammarion.

United Nations Convention to Combat Desertification (UNCCD) (2023). *The Great Green Wall.* Available at: https://thegreatgreenwall.org/ [Accessed 20 January 2023].

Warka Water (2022) *Warka Water. Every Drop Counts.* Available at: <https://warkawater.org/> [Accessed 20 January 2023].

Weller, J.R., Hoch, C. and Huang, C. (2017a) *Atlas for the End of the World.* Available at: <https://atlas-for-the-end-of-the-world.com/> [Accessed 20 January 2023].

Weller, J.R., Hoch, C. and Huang, C. (2017b) *Sea level Rise.* Available at: <https://atlas-for-the-end-of-the-world.com/world_maps/world_maps_sea_level_rise.html> [Accessed 20 January 2023].

Xu, C., Kohler, T.A., Lenton, T.M. and Scheffer, M. (2020) 'Future of the human climate niche.' *PNAS*, 117 (21), pp. 11350–11355.

Zhai, J., Wang, L., Liu, Y., Wang, C. and Mao, X. (2023) 'Assessing the effects of China's Three-North Shelter Forest Program over 40 years.' *Science of The Total Environment*, 857 (1), pp. 1–12.

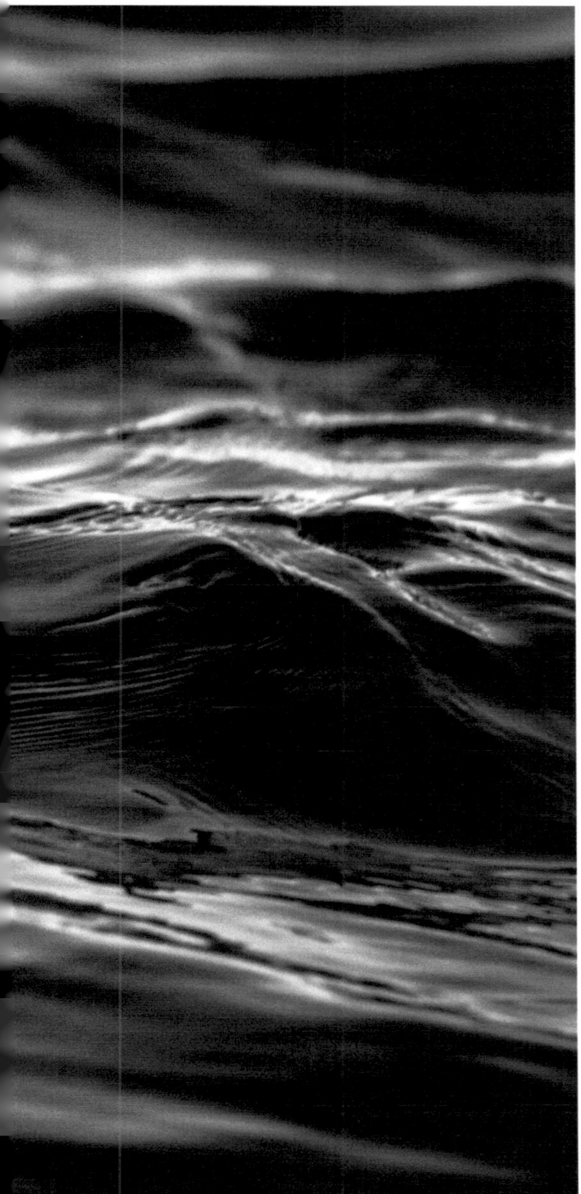

Part I

Liquid | Water

2 | Sponsland
A Collective Design Manifesto

Peter Veenstra

Design Project Data

Location
Province of Groningen | The Netherlands | Europe
Project Team
Municipality of Groningen | Province of Groningen | GRAS Foundation | Yttje Feddes | Peter Veenstra
Date
2020 | 2021
Coordinates
53° 13' 8.635" N | 6° 34' 4.825" E
Area
1.200 km²
Length
55 km
Investment
n.a.

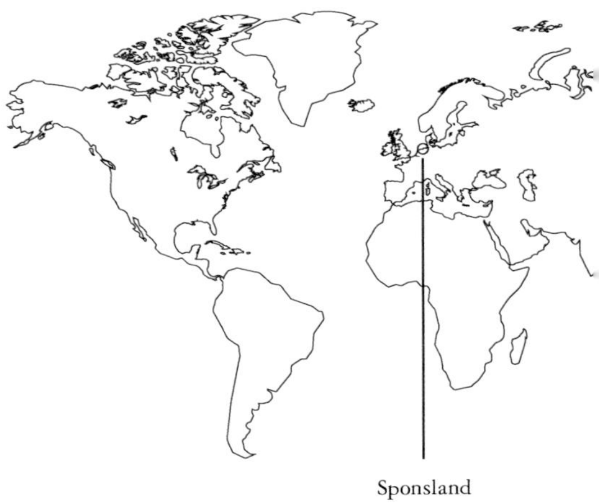

To accelerate the thinking and practice of climate adaptation in the Dutch province of Groningen, a design event was organized in 2020–2021. A total of eight design agencies were invited: Agence Ter, Bureau B+B, Flux, LIST, LAMA, MORE landscape with VE-R, Tredje Natur, and West8. They were asked to adopt an area and propose a visionary yet realistic future scenario using the necessary design adaptation as leverage for a holistic upgrade of the landscape. Each of the eight design areas is located within the boundaries of one river basin, crossing various landscape types. Together these areas form a colorful yet coherent perspective on what climate adaptation can do at a systemic level. Many climate solutions have been elaborated, including a 'ditch revolution,' a 'water battery,' an 'urban lagoon,' and an 'edible coast.' To inspire others involved in the spatial planning of climate-sensitive regions, the insights gained from the collective exercise have been extracted from their specific regional context and generalized into a manifesto (Ekamper et al, 2021a; 2021b) outlining ten principles for climate-adaptive design.

Introduction

Climate change is undoubtedly humanity's greatest challenge. To prevent global warming beyond critical values, society must adapt in all its facets. However, there is no guarantee that humanity will be able to slow down and eventually stop global warming. Even if it succeeds, the effects of global warming already taking place will be severe. The natural systems to which our land use and daily lives are geared with high precision are changing and displaying increasingly unpredictable behavior. This calls for adaptation to an uncertain future.

Design research can help to find new pathways for the future. The images produced by such research may even become self-fulfilling prophecies: Ideas so appealing that they will eventually be adopted because people will want to start living them.

This article describes such a design study: The design manifestation of Sponsland. It was named after China's Sponge City policy (Chan et al, 2018, pp. 772–778), which aims to increase city water storage capacity. The exhibition of Sponsland was one of the events featured at the Climate Adaptation Summit (CAS2021) (Bapna and Chan, 2021) that took place in the Netherlands in January 2021. The summit aimed to exchange global knowledge on climate adaptation and develop an action agenda. Sponsland, an initiative of the Municipality of Groningen and the Province of Groningen, follows the rich tradition of Groningen architecture events, such as Wonen in Stadshart (2017), BouwJong! (2010), Intense Laagbouw (2008), De Intense Stad (2004), Blue Moon (2001), A Star is Born (1996), and What a Wonderful World (1990). With Sponsland, the focus was on landscape and public space for the first time.

Sponsland focused on how climate change adaptation can be an engine for positive change. Although the challenge of climate change may be our most significant, it is not the only one. Other factors include growing social inequality and declining biodiversity. The shrinking countryside and the consequences of natural gas extraction also challenge Groningen. Thus, climate adaptation has been used to provide solutions across many other fronts through spatial projects. By preparing for the consequences of a changing climate now—from floods to heat and drought—we can avoid extremely high costs for future generations while also making our environment more sustainable. Doing so can reduce temperature rise, restore ecosystems, and establish a sustainable economy that does not deplete the earth's resources.

Eight design companies (Agence Ter, Bureau B+B, Flux, LIST, LAMA, MORE landscape with VE-R, Tredje Natur, and West8) examined the future of Groningen's landscape through design research. Their goal was to develop feasible plans—plans that could be implemented. The longer we delay making necessary climate adjustments, the higher the social, environmental and financial costs will be. Action taken early will also allow other improvements to be realized with that leverage. Although numerous visionary landscape plans have already been developed for the Groningen region over recent decades, their impact on reality was limited. The value of another visionary plan without action perspectives would be limited. For this reason, the design tasks were formulated with input from local parties. Moreover, all involved designers linked their vision to concrete proposals for interventions.

Together, the eight designs provide a spatial perspective for the region. Sponsland is presented not only as a collection of solutions but also as a coherent territory with a distinct character that one could visit in the future. The design process revealed what this world could look like; a diverse landscape where fresh water, salt water, agriculture, and nature merge and provide a rich array of agricultural products. A slow, spongy landscape with much more water: Buffer water, drinking water, bathing water, play water, boating water, and especially lots of reeds.

Following a brief description of the design studies, this article explores a more in-depth study of the lessons learned and how the impact of design research can be maximized.

Sponsland

From the Drents Plateau to the North Sea, Sponsland shows a cross-section of the Groningen landscape. It roughly encompasses the basins of the rivers Hunze and Aa, converging in the Reitdiep. Here, we find most soil types and water conditions along the Dutch, German, and Danish coasts. This makes for an ideal study area since many solutions from this region can also be applied elsewhere. Notably, four of the eight design sites are located in urban areas and four are located in rural areas.

For each site, the companies were asked to anticipate how the climate will develop over the next 100 years. This raises several critical questions: Can we keep the coastal landscape livable and productive with advancing salinization? Is it possible to restore biodiversity and close local cycles due to changing food production? What is the most effective strategy to retain valuable fresh water, including the preservation of vulnerable peat soils? How can we keep cities cool, dry, and pleasant? Can we tie this to a new urban economy and sustainable mobility?

Since all eight sites were situated along lengthy landscape and urban lines, addressing local characteristics and tasks was possible while retaining a greater view of the city and landscape. Together, the plans provide a future perspective on the entire region. Each agency learned from the other through joint working sessions while emphatically pursuing their interests. Thus, the Sponsland provides a rich palette with several common threads.

The collective future perspective on the region envisions a world where soil functions as a sponge and urban systems are not geared for acceleration, but rather for deceleration. Moreover, the vitality of historical heritage is being reinvented through climate adaptation. Furthermore, agriculture and nature are increasingly mixed. While the city is being used as a stage and testing ground, new deals are emerging between farmers, nature organizations, water boards, and residents to leverage each other's investments in rural areas.

In delta areas such as the Netherlands, climate change conversations often focus on flooding hazards (De Bruin et al, 2009; Kwadijk et al, 2010; Uittenbroek, Janssen-Jansen and Runhaar, 2013; Schultz van Haegen and Wieriks, 2015; Bloemen, Van der Steen and Van der Wal, 2018). However, insights from the present study reveal that the most significant issue of the future would not be excess water but a lack of high-quality fresh water (Ekamper et al, 2021a; 2021b). As such, many of the projects reflect this potential future context. New approaches to water are bound to lead to new agricultural reforms in which nature-inclusive circular agriculture—influenced by dynamic water levels and/or salinization—will create brand-new regional menus. In line with this, the Dutch Ministry of Agriculture, Nature, and Food Quality presented their vision for the future of agriculture: 'Agriculture, Nature, and Food: Valuable and Connected' (LNV and Ministerie van Landbouw, 2018). Agriculture reform will need to go hand in hand with adapting our diet.

The Eight Designs

Rather than describing each design study in detail, this article offers a bird's eye view from south to north. A journey through the future landscape of Groningen—the future we should desire.

Hondsrug is the highest point in the planning area. It is an elongated sand ridge south of Groningen City that hosts many estates and luxury residential areas. Agence Ter envisioned it as a woodland park, both a nature reserve and a living environment. The natural forest creeps onto the squares and streets, eliminating heat islands and improving water retention. The woodland park combines both agriculture and forestry.

The Hunzevallei lies parallel to the Hondsrug, where fresh water must be stored to prevent peat oxidation, obtain more drinking water, and replenish the natural water system downstream. To this end, Bureau B+B designed a 'water battery,' a new water system that retains water on the higher plateau and the valley slopes, thus creating new recreationally attractive wetlands in the valley.

In the historic city center of Groningen, which lies at the extreme tip of the Hondsrug, the architects from LIST suggested that the outdoor space design should be viewed as something other than an infrastructure task.

They used a park design including a succession of places and atmospheres, with each one avoiding heat and waterlogging differently. By adding 100 water elements, they propose changing the role of water and adding a new layer of complexity.

The inner city is connected to the surrounding countryside by several axes. Along one of those axes, the so-called Sugar Axis, West8 proposed reclaiming urban space by reducing car traffic and using this space for cooling, water storage, and sustainable mobility. This is enabled by a mobility hub combined with a food hub, thereby connecting rural food supply with urban demand.

A canal forms a second axis, with obsolete, car-oriented business parks along it. Tredje Natur developed a clever step-by-step plan to turn this area inside out, transforming the business parks into mixed live/work neighborhoods facing the water. A climate-adaptive water system, including a large freshwater reservoir along the canal, structures the neighborhood. This goes hand in hand with people-oriented mobility and regenerative biodiversity to counter the urban heat island effect.

In the north of Groningen, the Reitdiep flows through a centuries-old agricultural landscape at risk of salinization and aridification. The landscape office MORE proposed a combination of pre-modern and civil engineering interventions to make the landscape more resilient. On the one hand, a 'ditch revolution' is proposed based on the notion that with many small new ditches, more water can be captured and the thin clay layer can retain more moisture during dry summers. On the other hand, the separation of clean river water and polluted canal water is proposed, representing an unbundling of the river and canal system to dramatically increase natural values.

A key concern on the Wadden coast is the contrast between dynamic nature and industrial agriculture since salinization is encroaching and sea level rise threatens the tidal nature of the Wadden Sea. Flux proposed introducing new ambiguity to resolve this tension. Their proposal for an edible Wadden Sea coast embraces nature as a place for food production. It takes salinization as an opportunity for developing a short-cycle agricultural area with incredible diversity. In this proposal, local food does not represent a reduction but rather an increase in choices.

While all previous plans are interconnected and influence each other in terms of water management, the island of Schiermonnikoog stands alone. The greatest future task here is conserving and securing fresh water. To achieve this, Bureau LAMA created a water cycle that refills the dunes' freshwater bubbles. Here, a mixture of dynamic nature management and food production is also created, yielding enough to feed residents and tourists year-round.

In all eight cases, climate adaptation can be solved in an attractive and technically realistic manner. Together, the projects illustrate what tantalizing

Figure 2.1 | Hunzedal is turned into a water battery for downstream areas by Bureau B+B. Image by Bureau B+B urbanism and landscape architecture/MARTA pictures.

Legend

EAST-WEST AXIS
1a | Praediniussingel
1b | Emmaplein
1c | Ubbo Emmiussingel
1d | Hereplein
1e | Heresingel
1f | Radesingel
1g | Blauwe Bankjes
2a | Kleine der A
2b | Museumtuin
2c | Reitemakersrijge
2d | Minervaplein
2e | Gedempte Zuiderdiep
2f | Gedempte Kattendiep
3a | Akerkhof
3b | Vismarkt
3c | Grote Markt
3d | Forum
3e | Martinikerkhof
3f | Prinsentuin

POCKET SQUARES
4 | Northern Maritime Museum
5 | Munnekeholm courtyards
6 | Minerva Academy courtyard
7 | Pelstergasthuis
8 | Herestraat
9 | Herestraat square
10 | Rademarkt
11 | Pepergasthuis
12 | Achter de Muur square
13 | Poelestraat squares
14 | Sint Jans courtyard
15 | Visserbrug square
16 | Lopendediep square
17 | Harmonie Complex
18 | Turftorenstraat square
19 | Grote Kromme Elleborg
20 | Rijkuniversiteit square
21 | Mercado
22 | Het Klooster square
23 | Kwinkenplein

SATELLITES
24 | Station square - H.N. Werkmanbrug
25 | Damsterplein
26 | Hanzeplein
27 | Guyotplein - Ossenmarkt
28 | Westerkade - Sluiskade

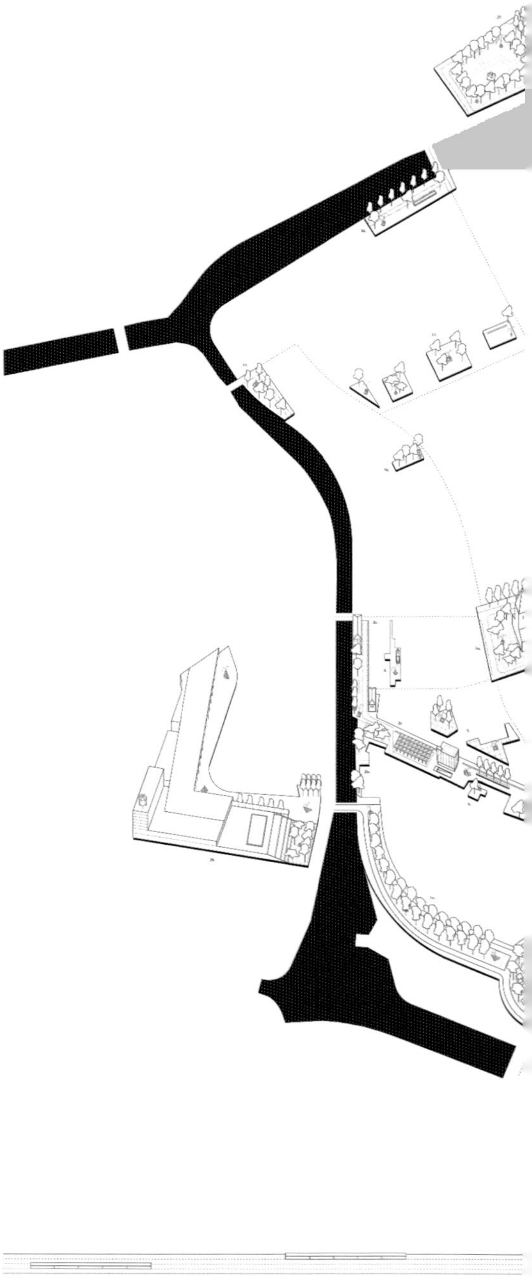

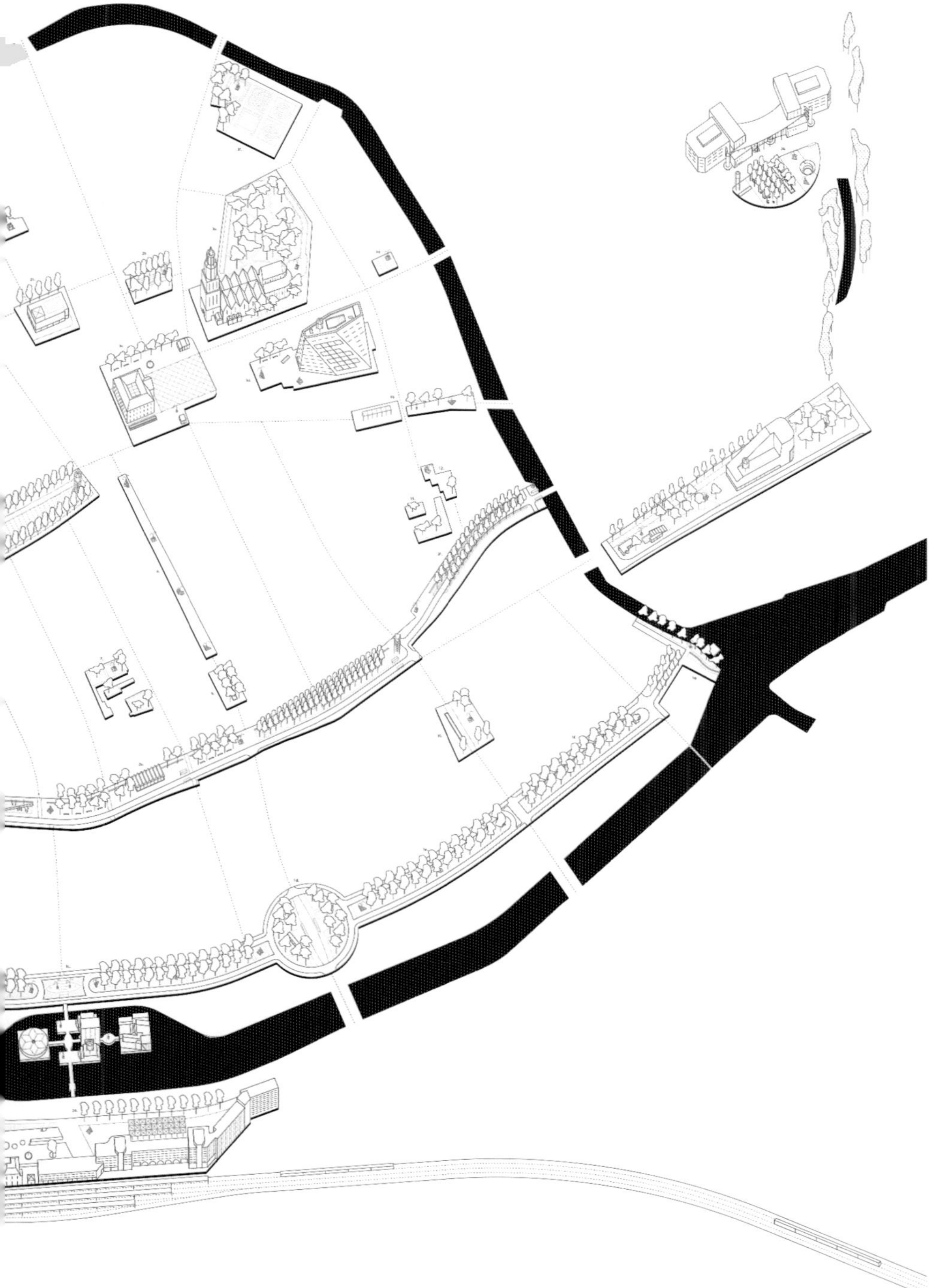

Figure 2.2 | 100 water elements for the historic inner city of Groningen. Image by LIST.

Figure 2.3 | An urban lagoon and a second dike form a contemporary blue rampart. Image by Tredje Natur.

new variations of climate adaptation can bring to one region. Thanks to the involvement of local actors and stakeholders, they are all preliminary studies that can serve as the prelude to an actual project.

The Sponsland Manifesto

The design studies and mutual discussions resulted in insights and shared convictions being developed. The idea of writing a manifesto (Ekamper et al, 2021a; 2021b) emerged as a way to translate these insights into more widely applicable principles for climate-adaptive design. The manifesto outlines ten principles for an effective approach to climate-adaptive design in public spaces and landscapes. The principles address both the design process as well as the spatial planning process and substantive design principles. Brief descriptions of these processes and principles are presented as follows.

Draw the Roadmap

The first point relates to how fundamental uncertainty should be dealt with. Climate scenarios show a broad spectrum of possible futures. With the addition of the volatility of our economy and short-term crises, it becomes clear that reliance on rock-solid long-term projections is no longer possible. This new reality requires designers to develop new strategies. We must draw the roadmap while we are on the journey. This roadmap has no objective narrative but rather an infectious vision and a moral appeal to help create a better future. The roadmap should list short- and long-term objectives side by side. Since global warming and rising sea levels will continue for some time, radical, long-term system changes may be necessary for vulnerable areas, such as the Dutch Delta. In the near future, a gradual adaptation of the current situation will suffice. Both Plan A (optimization of the existing) and Plan B (radical change) are equally important and should be developed simultaneously. However, the prospect of radical change in the distant future should not paralyze current actions.

Cultivate the Conflict

Adapting to climate change places new demands on our landscapes and public spaces. As a result, it will only be possible to satisfy some of these demands. Notably, it is not possible to find a win-win solution to every problem. Thus, politicians, policymakers, and designers must find a way to avoid the resistance and anger of residents. Change will not occur rapidly enough if plans are too safe or vague. It is essential to move beyond pretty pictures and be clear about the problematic knots that must be cut now and in the future. History has shown that conflict can be an engine for change. By cultivating conflict as a productive force, new solutions can emerge.

Develop New Standards

Climate change is both a physical and scientific challenge. Testing, evaluation, and knowledge sharing are crucial for accelerating the collective learning curve. This helps us avoid mistakes that we cannot afford to make in the future. Designing for climate change requires the use of new terminology, the adoption of standard solutions, and the development of new assessment systems. In the architectural field, sustainability labels and certificates are common; curiously, they are rare in public spaces and landscapes. However, quality labels may lead to a 'tick-box' mentality that discourages critical thinking. Despite this, there can be no doubt regarding the positive impact of quality labels since solidifying soft value is essential in the long run. For the significant transitions we face, these labels are a powerful tool.

Extract the Invention

Inspiring climate-adaptive designs can serve as examples worldwide, making an impact that transcends the place it was initially designed for. The goal of every designer should be to make a positive global impact with their work. This is why it is essential to be visible and develop ideas that are reproducible, scalable, and feasible. Therefore, the ideas must possess a generic quality. Paradoxically, truly sustainable landscape design is site-specific: Embedded in an area's physical, social, and historical context. Thus, the challenge involves developing proposals that simultaneously possess generic value and integral customization. Distilling copyable inventions from a site-specific project makes it relevant to other places.

Make Beautiful Machines

To transition to a sustainable future, our society must undergo a physical, social, and economic transformation. The development of a new, sustainable lifestyle is essential. This cannot be forced upon us; instead, we must be seduced. Just as the success of Apple and Coca-Cola cannot be separated from the quality of their product design, the success of the climate transition will depend on the form we give it. Aesthetics have a crucial role to play and deserve a firm place on the agenda again. We should not only talk about it; we should act upon it.

Notably, there is a lot at stake. If we do not invest in beauty, the transition could easily result in an uninspiring layer of 'generic greenery' that destroys local identity as well as cultural-historical significance. Thus, it is necessary to develop a new form of language. Engineers and designers should work together to develop beautiful nature-based water storage and heat stress systems.

Challenge History

If we consider climate mitigation and adaptation together, the transformation of many areas will go beyond subtle adaptation. It will be more like a total makeover. Change is especially difficult in culturally and historically valuable areas. When one is surrounded by beauty, global climate problems seem far away. Nonetheless, many culturally and historically significant areas must be adapted. This transition may be the most difficult task we face since it will be necessary for designers to challenge history. They will have to wrestle with it to understand its ability to change. A common strategy is to turn to the past for guidance. A romantic return to traditions, techniques, and old landscape patterns is tempting as a counter-reaction to contemporary dynamics. Yet, this will not be sufficient in the face of climate change. Intelligent, playful reinterpretations of the past are needed, as is the integration of modern technology and bold contemporary interventions.

Start With the Soil

Although climate change primarily manifests itself in the atmosphere and water, giving the soil equal consideration as a third element is appropriate. A climate-adaptive landscape starts with the soil. In the positive scenario, soils absorb changing rainfall patterns; in the adverse scenario, soils degrade as a result. Depending on the land use, soil can either increase or decrease atmospheric carbon concentrations.

The changing climate calls for restoring the balance between soil, water, and the atmosphere. Currently, that balance still needs to be achieved. The urban subsoil is smothered under the weight of motorized traffic. In rural areas, soil is being destroyed by industrialized agriculture. It is essential to approach the soil as a whole. What can the soil handle? How does it behave? What are its limitations? Developing integer terroirs—in which soil, tillage, and crops are interrelated—is required for landscape resilience.

Use All Tones of Blue

The most significant impacts of climate change are water-related: Sea level rise, heavy rainfall, and drought. The need to solve these quantitative water problems provides an opportunity to simultaneously improve water quality. The Netherlands is competent at water management, although this mainly focuses on water quantity. Notably, water quality should receive much more attention. Too often, clean water is wasted while polluted water has the chance to spread. Rainwater, groundwater, seawater, canal water, agricultural water, seepage water, sewage, and gray water; just as the Inuit have countless words for *snow*, we must learn to better distinguish and use the different water qualities. For more bathing water and biodiversity, clean water needs to be retained as long as possible. New nature can help purify water. In addition, new civil engineering structures are required to separate, buffer, and purify the various water flows. Additionally, agriculture can respond to local differences in the water system with crops that are resistant to salt, drought, and floods.

Reconcile Food and Nature

The post-war modernization and scaling-up of agriculture have led to polarization in the landscape. Agriculture and nature have become antagonistic to one another. They keep each other in a stranglehold while they are both in trouble. To escape from this damaging polarization, it is better to seek nature's productive potential and agriculture's natural potential. If we develop such a common interest, a healthy agricultural economy and resilient ecology can go very well together.

Currently, land use in rural areas is tied up in regulations and agreements with a limited number of players: farmers, water boards, nature organizations, water companies, and municipalities. It is time to make new deals, adapt existing regulations, and incorporate new players. What could those deals look like? That is where designers come in. They can help create new perspectives and engage and inspire new (and old) players.

Design for Slowdown

Over the last century, traffic design has dominated the design of urban public spaces. Even with traffic engineers paying more attention to slowing traffic and improving public transport, the emphasis remains on optimal traffic flow and efficiency. If priority is given to rainwater infiltration, cooling, and improving the urban climate in the design of public spaces, traffic efficiency becomes less critical. Traffic flows will slow down. The slow-green city can become an atmospheric place full of beauty, diversity, and surprise. It may take a few extra minutes, but getting around this city will be much more pleasant. Gone are the days when green urban ambitions were mainly expressed in defined green spaces. Free to Bruce Mau, landscape design should no longer focus on 'islands of good behavior in a sea of stupidity' (Ekamper et al, p. 20). The entire city should be approached as one big park: a place for nature that is optimized for human use.

Conclusions

The urgency, consequences, and challenges of climate change are not new. Ever since the 1970s, it has been clear that there are limits to our growth-oriented society and that the consequences of unbridled growth will sooner or later lead to disruption. In the Netherlands alone, this awareness has led to an extensive stream of design studies that call for a radical change in how we deal with nature, landscapes, and water. These studies were partially sponsored by the Creative Industries Fund NL, commissioned by the Board of Government Advisors and by the International Architecture Biennial Rotterdam (Dunn et al, 2017, pp. 71–79), or the outcome of design competitions like the EO Wijersprijsvraag. Over the years, these plans have not become obsolete; instead, they have only become more relevant. Yet there were many reasons for the failure of many plans to come to fruition. Procrastination, the nuance of the problem, and the system's rigidity are just a few of them.

A positive exception to this picture is Plan Ooievaar (De Bruin et al, 1987), a development vision for the Dutch river area resulting from the first EO Wijers competition. Even though the plan was not realized one-to-one, it did introduce a novel way of thinking and led to thousands of hectares of new natural features along the Dutch rivers. Plan Ooievaar shows the potential of an extensive and integrated development vision. Meanwhile, the vast collection of unimplemented plans raises the question of how much energy should still be put into developing new plans and visions. Should a much greater focus be placed on implementation? Or at least on linking the free space offered by a design manifestation like Sponsland to practice?

It is precisely this insight that made Sponsland strive to be more than a free study. Despite the freedom given to the teams, they studied concrete tasks. The choice was to associate *real* clients and problem owners with each location. The purpose of this project was to demonstrate what can be accomplished by landscape architects. Most of the value comes from linking tasks and opportunities. Adaptation to climate change cannot be achieved through sectoral solutions. There was also a challenge in emphasizing both the urgency and the opportunities related to climate change. This implies the need for a balance between a grave story about the seriousness and scale of the problems on the one hand and a positive story that illustrates the potential beauty and strength of transitions on the other. Additionally, a balance was sought between climate solutions addressed at a large, systemic level and smaller interventions that can already prove their value—even if small—in the short term.

Despite the positive effects of involving problem owners, gathering design talent, and working on concrete tasks and broader messages, there remains room for improvement in the shift to practice. The key to sustained success is follow-up and finding effective ways to engage problem owners, municipalities, provinces, water boards, residents, and interest groups over the long term. To prevent falling back into old habits, the design exhibition should not be a festive ending to the process but the beginning of a serious next project phase: Implementation. The formula of the Groningen design manifestations works well for tackling complex issues in spatial planning and could—in its improved form—be successfully applied to other places. While avoiding the climate catastrophe has changed from a long-term problem into a short-term problem, many regions and cities would benefit from making the manifestation approach a regular part of daily practice.

Acknowledgments

This article was written with the help of Peter-Michiel Schaap, director of GRAS, and Yttje Feddes, co-curator of the design manifestation the article describes.

References

Bapna, M. and Chan, C. (2021) *Outcomes and Next Steps from the Climate Adaptation Summit.* Available at: <https://www.wri.org/insights/outcomes-and-next-steps-climate-adaptation-summit> [Accessed 29 December 2022].

Bloemen, P., Van Der Steen, M. and Van Der Wal, Z. (2018) 'Designing a century ahead: climate change adaptation in the Dutch Delta.' *Policy and Society*, 38 (1), pp. 58–76.

Chan, F., Griffiths, J.A., Higgitt, D., Xu, S., Zhu, F., Tang, Y., Xu, Y. and Thorne, C.R. (2018) '"Sponge City" in China—A breakthrough of planning and flood risk management in the urban context.' *Land Use Policy*, 76, pp. 772–778.

De Bruin, K., Dellink, R.B., Ruijs, A., Bolwidt, L., Van Buuren, A., Graveland, J., De Groot, R. S., Kuikman, P.J., Reinhard, S., Roetter, R. P., Tassone, V.C., Verhagen, A. and Van Ierland, E.C. (2009) 'Adapting to climate change in The Netherlands: an inventory of climate adaptation options and ranking of alternatives.' *Climatic Change*, 95, pp. 23–45.

De Bruin, D., Hamhuis, D., Van Nieuwenhuijze, L., Overmars, W., Sijmons, D. and Vera, F. (1987) *Ooievaar: De Toekomst van Het Rivierengebied.* Arnhem, Stichting Gelderse Milieufederatie.

Dunn, G., Brown, R.R., Bos, J.J. and Bakker, K. (2017) 'The role of science-policy interface in sustainable urban water transitions: Lessons from Rotterdam.' *Environmental Science & Policy*, 72, pp. 71–79.

Ekamper, T., Feddes, Y., Schaap, P.M. and Veenstra, P. (2021a) *Sponsland, Reis door het landschap van de toekomst.* Groningen, GRAS.

Ekamper, T., Feddes, Y., Schaap, P.M. and Veenstra, P. (2021b) *Manifesto: 10 Principles for Climate Adaptive Design.* Groningen, GRAS. Available at: <https://sponsland.nl/en/manifest/> [Accessed 18 December 2022].

Kwadijk, J.C.J., Haasnoot, M., Mulder, J.P.M., Hoogvliet, M.M.C., Jeuken, A.B.M., Van der Krogt, R.A.A., Van Oostrom, N.G.C., Schelfhout, H.A., Van Velzen, E.H., Van Waveren, H. and De Wit, M.J.M. (2010) 'Using adaptation tipping points to prepare for climate change and sea level rise: a case study in the Netherlands.' *Wiley Interdisciplinary Reviews*, 1 (5), pp. 729–740.

Ministry of Agriculture, Nature and Food Quality of the Netherlands (2018) *Agriculture, Nature and Food: Valuable and Connected. The Netherlands as a Leader in Circular Agriculture.* Available at: <https://edepot.wur.nl/509271> [Accessed 18 December 2022].

Schultz van Haegen, M. and Wieriks, K. (2015) 'The Deltaplan revisited: changing perspectives in the Netherlands' flood risk reduction philosophy.' *Water Policy*, 17 (S1), pp. 41–57.

Uittenbroek, C.J., Janssen-Jansen, L.B. and Runhaar, H.A.C. (2013) 'Mainstreaming climate adaptation into urban planning: overcoming barriers, seizing opportunities and evaluating the results in two Dutch case studies.' *Regional Environmental Change*, 13 (2), pp. 399–411.

3 | Let the River Run

The Aire Waterway as a (Co)Designer

Georges Descombes

Design Project Data

Location
Geneva | Switzerland | Europe
Project Team
Superpositions
Date
2000 | 2022
Coordinates
46° 10' 24" N | 6° 05' 03" E
Area
n.a.
Length
5 km
Investment
60 million euro

Aire River Project

The Aire River project is an innovative process-based design, with its lozenge diagram of excavation being both a spectacular accelerator of the self-morphogenesis of the river and its sediment reservoir. A decisive shift in the conception of stream restoration is the abandonment of designs based on the search for a form directly given to a new riverbed (form-based design) and the adoption of designs based on natural processes (process-based design). Such designs seek physical devices for initiating a self-morphogenesis of the river.
Four criteria have been proposed to evaluate the quality of such designs:

Space given to the river (divagation space or freedom space, which returns the spatial field of stream processes lost by human actions).

Energy (evaluating the design actions based on natural energy in the realization while minimizing the external use of mechanical energy).

Materials (use of existing sediments to facilitate the evolution of the new bed).

Time (the amount required to reach the objective of achieving a new morphology).

A topographical imagination

The Aire River flows south of Geneva through valleys historically devoted to farming. From the late nineteenth century to the 1940s, the river was progressively canalized to control flooding. In 2001, the State of Geneva opened a restricted competition to return the river to a more natural form. The general implicit idea of the competition program was to restore the river to its original shape and setting by destroying the canal and redirecting the flows into the former meanders. The primary justification for this undeclared demand was the distaste for the unnaturally *straight canal line*. It represented a return based on the well-anchored conviction that it is necessary and possible to return to a state of pure nature. A search for a *lost paradise* was, in fact, the narrative behind all of this.

Instead of this suggested solution, a plan was proposed that combined the existing canal, transformed, with a vast new parallel *divagation space* for the river. In the process, the canal becomes the *pointer* (indication) for the transformations, a reference line that allows the visitor to understand and feel the *before* and *after*. A *becoming* that superimposes both situations (Descombes, 2018).

Any such project can shift given situations and may not represent a return to a former so-called *historical*, *original* state of things. In the 1980s, this *design attitude* was put forward at the School of Architecture at the University of Geneva under the illuminating influences of André Corboz (Corboz, 2001) and Bernardo Secchi. In the beginning, possible applications of painting restoration methods were tested on architectural designs. Then, enlarged and applied to the whole territory, this progressively gave birth to what was called a *topographical imagination* (Rosenberg, 2001; Marot, 2003), where the description of situations and the proposed transformations were carefully thought out and controlled. This approach became the basis on which selected points of radical modifications or intensifications were chosen. In this patient examination of the context for the interventions, historical documents, plans, and images were equally considered along with existing *on-site* observations and measurements. Ideas and imagination lead to unrevealed potential. The philosopher Deleuze, in the conference, asks *Qu'est que l'acte de création?* ('What is the act of creation?').

> 'What does it mean to have an idea? (...) Ideas, we must treat them as a kind of potential, but already engaged in a particular mode of expression and inescapable of this mode of expression.'
>
> (Deleuze, 1987)

Keeping with this *exploratory character*, the Aire River project aims to reverse what was implicitly planned in the competition, which represents mere negations—the negation of the territory, its history, and the gradual sedimentation of its formation and transformation through the ages. The design opposes the destruction of all relationships between the cultural and architectural traces and the overall morphology of their context. The river is the (co)designer of this ever-evolving landscape:

> 'As reconceived by the Superpositions design, the landscape becomes a framework for natural processes and human recreation. Each aspect of the design led to a form addressing a programmatic or scientific concern: The management of water, the growth of plant communities existing and introduced, and motor vehicle and pedestrian circulation. Water flow, rain, snow, sand, gravel, wind, vegetation, and temperature will ultimately determine the ever-changing landscape of and around the Aire. A partnership between river and landscape architect results, or as the Superpositions designers phrased it: the river became their co-designer.'
>
> (Treib, 2018, p. 253)

L'Aire, an Open-Air Laboratory

The transdisciplinary composition of the research team, required by the very nature of a river restoration project, provoked an element of experimentation since the onset of the studies. Biologists, environmentalists, hydraulic and structural engineers, and landscape architects, from their specific knowledge and surpassing them, together elaborated the project's hypothesis and invented the means to verify it.

The project proposes a research path incorporating urgent ecological shifts into a larger cultural change. As described above, the complex organization of the design associates the new river space and a linear series of gardens in the former canal. In reality, the entire design becomes *one linear river* garden. The garden is understood as one part of the trilogy of three natures (Hunt, 2000)—the garden is a place of questioning as much as a place of pleasure. Facing the entire catchment, the original morphology of the mountains, and the traces of human modifications, this long *river garden* organizes the situations, views, confrontations, presences, and materials, aiming to introduce questions, worries, and hopes into this fragile and precious territory. The necessary *calm and interiority*, without which there is no real garden, yields organized sequences of differentiated places and paths, thus allowing a reasonable distribution of people and movements.

The *footprint* of the canal is a key device for building this new organization, a permanent trace that limits and frames a succession of situations and introduces a complex temporality.

In 1998, I attempted to incorporate a few traces of the former state of the site that remained perceptible in the design of the Bijlmer Memorial in Amsterdam, proposed after an airplane crashed into a housing complex.

Figure 3.1 | The aerial photo shows the Aire River and the canal from above. The original canal has been covered up, and the river runs in the lozenge design riverbed. Photograph by Fabio Chironi.

The footprint of the destroyed building was retained as part of a new public garden—a mix of evocation, a memorial, but also hope, growing in the garden among what has disappeared and what remains.

In the Aire River project along the former canal, a certain number of *brutalist/shocking* architectural devices have been introduced to renew the attention of visitors and initiate a new understanding of these places through a new feeling, an emotional impact opening an *involuntary memory* process capable of short-circuiting time and space. Such *provocations* divert expectations and wake up deeply covered, forgotten sensations. A concrete slab cantilevered above the water incites you to come nearer and hear, smell, and touch the flow.

> 'The canal garden, a hybrid of rural and urban typologies, provides a new public space that restructures the fragmented territory. The use of landscape as a large-scale planning device recalls earlier Swiss experiments in the making of rural open space. A regional green network was already imagined for the Canton of Geneva in the visionary 1936 master plan for the Canton proposed by Maurice Braillard (…). But while the Aire canal garden promenade draws on this tradition, its attitude to landscape is radically different.'
>
> (Rosenberg, 2018, p. 56)

Diagram

As for the *design and construction* of the riverbed itself, two successive phases of realization have produced two different manners. As Deleuze suggests, we imagined a trace rather than a defined form.

> 'Something, a trace, a stain, which opposes the always threatening cliché, to a form too soon arrived.'
>
> (Deleuze, 1981, p. 67)

Obviously, *to design a river* is an intriguing term since alluvial rivers design their own channels and floodplains freely and perfectly. Conscious of the pointless effort to design and build a *fixed* riverbed—an approach often taken by engineers or landscape architects through *technical calculation* or *naturalistic mimicry* (Kondolf, 2006)—the surface layer of humus was excavated in the first experimental phase. Thereafter, the river flowed and did its job.

The result was a potentially perfect river design since the river flow gradually carved out a complex channel in the exposed substrate and established gravel bars through deposition.

However, of course, the river *designs* mainly when the massive fluid forces (which can easily exceed the equivalent of many days of bulldozer work;

Ciotti et al, 2021) can rearrange sediments and result in channel formation. Consequently, the process can—and certainly does—take time. However, to produce habitat *in a shorter time frame*, the relevant government agency (*Geneva Service de la Renaturation des Cours d'Eau*) urged to devise some *accelerating devices* to help the river more rapidly diversify its new bed.

Looking for an answer, the phenomenon of *percolation* (the act of a liquid passing through a porous material) was ultimately considered, and drawings and models were made (first with a bar of chocolate, a convenient pattern in Switzerland). Finally, this resulted in a diamond-shaped pattern, which opened a complex series of undetermined channels for the riverbed. Once again, these channels were excavated along the new river space by removing the soil's superficial layers while maintaining precise control over the longitudinal profiles of the riverbed. The dimensions of the *lozenges' islands* were configured to 'accept' the general sizes of the former meanders. The result, nearly a kilometer long for the first experimental segment, is spectacular and recalls the devices of land artists effecting clearly artificial interventions into a natural situation, followed by these being left to the mercy of natural forces. In fact, there is a familiarity with this artwork: It was *created by a deformation, a launching pattern* whose forms address the interplay between the river flow and the prepared terrain. In the words of Gilles Tiberghien:

> 'Just like Smithson's works, the Plaine de L'Aire project has a narrative that works with scales and organises particular moments for the visitor, sorts of time capsules that isolate them, integrate them into a new space-time without actually cutting them off from the world, the rhythm of which is made apparent to the visitor by the flow of the river, the sound of the water, the reflections of light, all of which at the same time permits a sort of cosmic daydream.
> The landscape of the Plaine de L'Aire has henceforth found its rhythm with this project, which has turned it into a work of art comparable with those often identified, whether or not correctly, as Land Art. But here, the alliance of technical imperatives and formal preoccupations seized in their dynamism has transformed the work of nature into a work of art.'
>
> (Tiberghien, 2018, p. 144)

Six months after the opening of this new river space, the results were beyond the most optimistic expectations: The river flows, transporting diverse materials, such as gravel, sand, silt, and wood. Over time, the geometrical matrix of lozenges is modified by successive floods, creating a complex succession of riffles and pools. The surprising rapidity of this phenomenon of self-morphogenesis may be explained by the dimensions of the channels in the diagram. A combination of *lateral resistance*, maintaining a flow power, and the simultaneous *fragilization* of the banks accelerates their

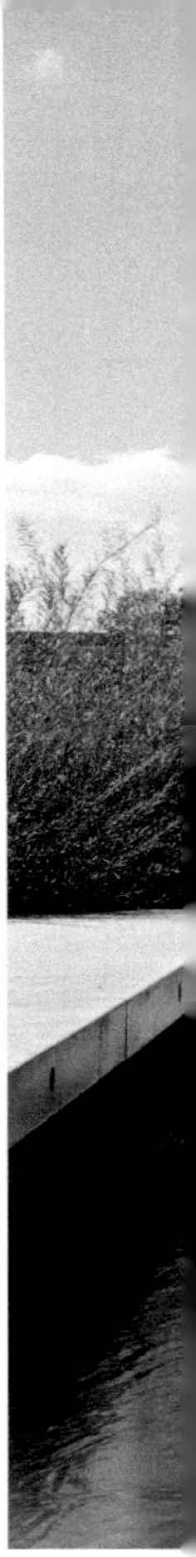

Figure 3.2 | Canal dikes.
Photograph by Jacques Berthet.

erosion. In contrast to conventional efforts to minimize erosion and conserve soils, the diagram is a system of *ravines* or *channels* designed to erode and transport sediment in reality.

A Paradox

We must accept this paradox: The more defined the grid given to the river, the more the river will be free to design itself. How, then, does one not think of the *urban grid*—a true example of a potentiality of evolution and diversification of urban forms—launched by an initial regularity? Dynamic forms and spatial construction were reworked and rebuilt over time, destroyed, and enlarged. And then, how does one remain unaffected by this parallel between two *impatiences*, those of forms too soon defined, congealed in the *grimaces* of towns or rivers? The diamond grid initiates an indirect action, allowing the river to shape and design independently:

> 'To act, therefore, in order for something to arise. But to act without doing everything.
> It is indeed necessary to act, but in such a way that something like a river comes about without our being able to determine theoretically, with exactitude, what its movements and its route are going to be. It is therefore a question of implementing an art of indirect action, of installing a set of artful systems that will supply the event with the conditions for its appearance and allow the "something," the river, to form, as it were, to take shape "on its own."
> Groupement Superpositions has invented the diamond system. This geometric structure concentrates in an exemplary manner the gesture of a project that unites two directions that could appear opposed (but actually are not): The direction of intervention and that of letting be. Here, the geometric grid is not destined to enclose, but to open: To open a space for possible shifts in the river's course, to open a space for future events. It is typically an expression of the will to make with (the river).'
>
> (Besse, 2018, p. 86)

Self-Organization

In different interviews, Edgar Morin approached the problem of *self-organization* by referring to Ilya Prigogine's structures at bifurcation points, where the ancient structure becomes unstable and new structures appear. This represents the birth of complexity. The concept of the bifurcation of self-organization is increasingly penetrating all of the sciences, passing from a world of certainty to a world of probability—a world in construction. How does one avoid being seduced by models of understanding reality that focus on a non-fixity of the elements and a generalized flux of materials? Was this not the experience in realizing our project, giving up any definitive design of the riverbed in the search for launching elements of self-organization?

The present shift in the models of design for river restoration from *form-based design*, in which an illusory form of the river is directly imposed (Kondolf, 2006), to *process-based design*, follows the same convictions, and the Aire design can easily be seen in this new perspective. Applying the four criteria for process-based restoration proposed by Ciotti et al (2021) (*new space given for the river, energy, materials, and time*), it is evident that the Aire River project responds well to the criteria (Kondolf, Descombes and Zingraff-Hamed, 2021). In particular, the Aire innovates in the fourth criterion, *time*, by spectacularly *accelerating* an autonomous morphogenesis of the new riverbed.

It is also true that the choice of the lozenges diagram for the earthworks provides a series of sediment reservoirs important to the new riverbed constitution and also represents an important economic aspect by avoiding the export of materials from the site works. The restoration approach implemented on the Aire thus allows the river to design its own channels. However, it continues to direct the river's energy along certain lines, an example of what has been termed 'prompted recovery' (Downs and Gregory, 2004).

An Experiment in Progress

The Aire River restoration now enters its fourth phase of realization, one more kilometer upstream to the French border. This phase once again takes up the lozenges diagram as the main design principle for the new riverbed, with new complexities due to the increased slope of the site, an expected volume of important sediments following the destruction of an existing dam, and the construction of a new pedestrian bridge designed by the Swiss engineer Jürg Conzett.

Notably, a particular context that has been decisive for the project's coherence must be highlighted. First, from an academic perspective, several project team members are involved in research programs on the landscape at the University of Geneva. Second, from a political perspective, the project has received continuous support from successive ministers on both sides of the political spectrum and, more importantly, the director of the *Geneva Service de la Renaturation des Cours d'Eau* has remained unchanged for more than 20 years of the project's realization.

It is undoubtedly too early to see the consequences of the design in the long term. A complete evaluation of the evolution of the Aire River is merited, mainly because it has been widely acclaimed. What justifies this affirmation and a possible generalization of this new way of designing a river space? The concept of the lozenge diagram, imitating earthworks and giving the river freedom to form its own bed, appears to have worked successfully. When walking along the river, one can see the results in the lack of severe floods and the acceleration of new riverbed formation (Descombes, 2021).

Figure 3.3 | Lozenges scheme.
Image by Superpositions.

Figure 3.4 | Lozenges excavation.
Photograph by Fabio Chironi.

Figure 3.5 | The image shows an aerial photo of the earthworks in preparation for the river flow. Photograph by Fabio Chironi.

Figure 3.6 | The photo pictures the river erosion as soon as water was introduced into the newly designed riverbed. Photograph by Superpositions.

Figure 3.7 | An aerial photo of the river erosion of the designed riverbed. Photograph by EasyToMap.

Nevertheless, one can now observe that the limits set on the new 'freedom space' for the river merit reconsideration since the river is eroding its banks such that it appears to be trying to regain the amplitude of its former meanders.

To allow this natural process and self-sizing to occur, stakeholders (notably farmers) will need to contribute by providing the river with an enlarged space. This extension of the project's limits is linked to an urgent change in agricultural production methods, in which the river Aire will be part of a new agroecological land organization instead of the present industrial agriculture system, which has dramatic consequences on the quantity and quality of the river's water. It is obvious that the true limits of a river restoration design should be its catchment, representing the only possibility for a chance to consider and master the ensemble of factors intervening in the river ecosystem. Nevertheless, one must accept the reality of acting under restricted conditions, with the hope that the actions taken will help to better understand the real problems while risking the potential for change.

In the case of the river Aire, the extensions and order of succession of the four realization phases have primarily been decided by circumstances and opportunities (urgent security measures against flooding and the availability of agricultural fields). The Aire River project is inscribed in a general concern for the dramatic changes in climate, which makes the question of water resources more crucial than ever. Under these conditions, the possibility of experimentation has nevertheless been real. The results are almost undoubtedly helpful advances in the process-based design framework.

How should we think of the Aire River's design? Merely as a form inscribed in the landscape, the 'courage' of a hypothesis (Corboz, 1997, p. 10).

References

Besse, J.M. (2018) 'The Diamond Paradigm.' In: Descombes, G., Descombes, J., Van Cauwenberghe, C., Correnti, V. and Gerber, F. eds. *Aire: The River and its Double*. Zurich, Park Books, pp. 83–90.

Ciotti, D.C., McKee, J., Pope, K.L., Kondolf, G.M. and Pollock, M.M. (2021) 'Process-based design criteria for restoring fluvial systems.' *Bioscience*, 7 (1), pp. 831–845.

Corboz, A. (1997) *"La" Recherche: Trois Apologues*. Genève, Droz.

Corboz, A. (2001) *Le Territoire Comme Palimpseste et Autres Essais*. Beançon, L'imprimeur.

Deleuze, G. (1981) *Francis Bacon Logique de la Sensation*. La Roche-sur-Yon, Editions de la Différence.

Deleuze, G. (1987) *Qu'Est Que l'Acte de Création?* Mardis de la Fondation Femis, 17th May 1987. Available at: https://www.lepeuplequimanque.org/acte-de-creation-gilles-deleuze.html [Accessed 20 July 2023].

Descombes, G. (2021) *Laisser Faire la Rivière*. Gollion, InFolio.

Descombes, G., Descombes, J., Van Cauwenberghe, C., Correnti, V. and Gerber, F. eds. (2018) *Aire: The River and its Double*. Zurich, Park Books.

Downs, P. and Gregory, K. (2004) *River Channel Management: Towards Sustainable Catchment Hydrosystems*. London, Routledge.

Eco, U. (1965) *L'Oeuvre ouverte*. Paris, Seuil.

Hunt, J. (2000) *Greater Perfections: The Practice of Garden Theory*. Philadelphia, University of Pennsylvania Press.

Kondolf, G.M. (2006) 'River restoration and meanders.' *Ecology and Society*, 11 (2), p. 42.

Kondolf, G.M., Descombes, G. and Zingraff-Hamed, A. (2021) 'Restoring dynamic fluvial processes in urban rivers: learning from the Aire and Isar Rivers.' *Landscape Architecture Frontiers*, 9 (4): pp. 10–27.

Marot, S. (2003) *Suburbanism and the Art of Memory*. London, Architectural Association.

Morin, E. (2016) *La Voie*. Paris, Pluriel.

Rosenberg, E. (2001) *Une imagination topographique*. Paris, Les Carnets du Paysage.

Rosenberg, E. (2018) 'The canal and the river.' In: Descombes, G., Descombes, J., Van Cauwenberghe, C., Correnti, V. and Gerber, F. eds. *Aire: The River and its Double*. Zurich, Park Books, pp. 53–61.

Tiberghien, G.A. (2018) 'Inventing the landscape of the Aire.' In: Descombes, G., Descombes, J., Van Cauwenberghe, C., Correnti, V. and Gerber, F. eds. *Aire: The River and its Double*. Zurich, Park Books, pp. 143–150.

Treib, M. (2018) 'A river as a co-designer.' In: Descombes, G., Descombes, J., Van Cauwenberghe, C., Correnti, V. and Gerber, F. eds. *Aire: The River and its Double*. Zurich, Park Books, pp. 245–253.

4 | Designing Alternatives to Megaprojects

Traditional Knowledge as Hope Against Seawalls in Japan

Annaclaudia Martini

Design Project Data

Location
Prefectures of Fukushima | Miyagi | Iwate | Japan | Asia
Project Team
Ministry of Infrastructure | Fukushima Prefecture | Miyagi Prefecture | Iwate Prefecture | Municipal entities | National university experts | Practitioners
Date
2013 | 2023 (ongoing)
Coordinates
39° 1' 39.81" N | 141° 37' 31.756" E
Area
n.a.
Length
400 km
Investment
1.35 trillion yen (roughly US $12.5 billion as of 2020; the project is not yet finished)

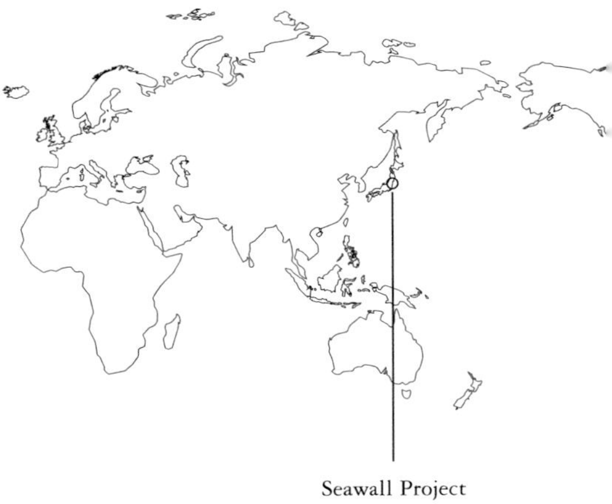

In a world where climate-related crises are becoming increasingly widespread, far-reaching, and unpredictable, it is important to *predict the unpredictable* and design not only structures but also futures that promote long-term resilience and hope. This chapter investigates the seawall and sea defense megaproject implemented by Japan after the March 11, 2011, triple disaster as well as the negotiations surrounding bottom-up alternatives to this project. In 2012, Japanese governmental institutions promoted the construction of 400 km of sea defense infrastructure in the coastal communities of Tohoku, northeastern Japan.

While this top-down technocratic approach to tsunami defense is broadly seen as necessary, there has been resistance from many in the affected communities since the seawalls in place before 2011 proved ineffective and lulled the communities into a false sense of security. Moreover, the massive walls would irrevocably damage the seabed's flora and fauna, and the megaproject does not consider the long-term socio-cultural aspirations and desires of the communities. However, coastal communities have hope: Through protests and negotiation, some changes have been implemented to complement sea defense systems with certain additions. These include changes to the sea defense structures, the use of forest walls, a rethinking of evacuation routes, and other historical practices that have once again become relevant since they are seen by the communities as more sustainable and in line with their ideals.

Introduction

In a world where climate-related crises are becoming increasingly widespread and unpredictable, it is important to *predict the unpredictable* and design not just better, more effective structures, but also futures that promote long-term resilience and hope. In this chapter, the discourses and negotiations around the construction of the 'seawalls and sea defenses megaproject' implemented in the Tohoku region of northeastern Japan are analyzed. These new structures and their designs have been contested by the residents of many coastal communities in Tohoku, which have proposed alternatives and mediations considered more in line with the future they desire for their communities.

These defenses are part of the recovery efforts following the 2011 triple disaster, a compound set of catastrophes in which a magnitude nine earthquake caused a tsunami reaching up to 38 m in height, which washed away most of the existing seawalls and breakwaters. Moreover, the combined effects of the earthquake and tsunami caused the nuclear meltdown of four of the six cores at the Fukushima Daiichi Nuclear Power Plant. Immediately after the disaster, which caused 20,000 casualties, governmental institutions at different levels promoted the construction of 400 km of sea defense infrastructure along the coast as part of the recovery package. While this top-down, technocratic approach to tsunami prevention has been seen as logical and sensible by Japanese governmental institutions, unexpected resistance from communities stemmed from the fact that: (1) the seawalls in place before 2011 proved ineffective at protecting the communities and lulled them in a false sense of security; (2) the massive new walls would irrevocably damage the seabed's flora and fauna and even exacerbate the consequences for the oceanic environment that climate change might cause; and (3) the megaproject does not consider the communities' long-term socio-cultural aspirations and desires. Although the construction of seawalls cannot be stopped, those who oppose them have hope since some communities have managed to negotiate changes in their design so that they consider impacts on both human and non-human elements. Moreover, discourses on alternative planning have spread, which have fostered a re-imagining of historical and traditional tsunami defense practices in use before the construction of seawalls.

Designing Sea Defenses After 2011

According to a survey from the Ministry of the Environment of Japan, 50% of the Japanese coastline had been altered by concrete infrastructures in 2001 (Ministry of the Environment of Japan, 2001), which cost an average of 1,500 billion yen per year (US $10 billion). However, on March 11, 2011, most sea defenses in place along the coast of Tohoku failed, being overtopped and destroyed. This elicited a rethinking and revising of Japan's design codes for sea defense structures (Raby et al, 2015).

The design of sea defenses involves a government design office—often assisted by engineers in the private sector—that produces a project following the Ministry of Land, Infrastructure and Transport's 2007 Technical Standards. These standards are based on The Disaster Basic Law, which was legislated in 1961. Thanks to the seawalls, damage from the 1960 Chilean Earthquake and Tsunami was avoided (Kimura, 2016), popularizing the idea that concrete sea defenses were sufficient to protect the Japanese coastline. Although seawall heights were increased in the 1980s, they could not withstand the unprecedented violence of the 2011 tsunami.

The disaster prompted revisions to Japan's prevention and mitigation policies. However, these revisions were found to be ambiguous, too vague, and difficult to apply (Aldrich, 2019). Two levels of tsunami were identified: Level 1, which includes tsunamis that have a relatively low height and high likelihood of occurrence, for which concrete defense structures are advised as the main prevention and protection tool; and Level 2, which refers to the largest tsunami possible in the area (Sato, 2015). Since this second type of tsunami cannot be mitigated through structural protection alone, sea defenses should be complemented with soft structures.

While these revisions show how the government considered both hard (seawalls and breakwaters) and soft countermeasures (forest walls and evacuation-based measures), soft measures were never envisioned as more than a corollary to infrastructural projects. When the project was first proposed in September 2011, it fell under the radar since residents were preoccupied with short-term recovery efforts (temporary housing, burying the bodies of victims, and cleaning up debris). However, as life slowly returned to normal, people started doubting the efficacy of the project and questioning the amount of time, money, and construction materials required (Kimura, 2016). Residents, environmentalists, and coastal scientists began to protest, concerned about the visual impact and degradation of the coastal ecosystem (Pinter et al, 2019). The government did not listen and pressed on with the construction of 400 km of massive seawalls and sea defenses, causing people to feel unheard, invisible, and abandoned.

The Issues With Japanese Sea Defenses: Designing for Enemies or Neighbors?

After 2011, rather than allowing local governments to use a participative decision-making process regarding the reconstruction, the central government defaulted to continuity (Aldrich, 2017). Grounding their reconstruction policies firmly in past experiences and the historical popularity of seawalls as a rationale to build even higher structures, standardized designs were implemented. However, as the world has become increasingly attentive to issues of environmental degradation and climate change, and in light of the complete failure of hard structures in 2011, seawalls are no longer appreciated as savvy prevention measures and are now considered myopic and dangerous. This is a known fact in the communities affected by this disaster, yet something that the Japanese government has failed to grasp. For many, the tsunami was not to blame; the seawalls were. Not considering this aspect in the design of the new sea defenses has been met with public outrage. Ultimately, Japan's 2011 debacle can be considered a failure of imagination. In failing to imagine what both a disaster and the most appropriate defense from it could look like in a time of climate change and rising sea levels, the nation was not prepared for the 2011 disaster. The massive number of casualties can also be traced back to an excess of faith in technology typical of the Japanese discourse: while praising nature and the environment in their literature and art, in reality, the country has been defined as a 'construction state' (Matanle, Littler and Slay, 2019) due to its readiness to deploy concrete to correct for an assumed dangerous and defective nature (Takezawa, 2016).

The seawall megaproject presents a set of economic, environmental, and social issues:

First, it creates a false sense of security. Relying solely on a technocratic infrastructure is one of the main issues highlighted by survivors of the 2011 disaster. Most people trusted the seawall to protect them and did not even attempt evacuation, or they first went home to collect personal belongings and check up on relatives.

Second, seawalls create a visual and physical barrier between the coast and the ocean. The main industry in coastal Tohoku is fishing, which is also the industry that is most impacted by such infrastructure. In an interview collected by the author, Naoya Takahashi, a fisherman in the town of Minamisanriku, explained that part of the 10-m-high wall was built right in front of his warehouse, blocking his view of the ocean and making the 15-m journey from there to his boat impossible without a car. Many towns also developed tourism as a way to regain economic revenue after the disaster, when fishing and farming were impossible. However, they now fear that the seawall could reduce the appeal of towns and thereby limit the number of tourists.

Third, because many households have now relocated to higher ground, high seawalls have become unnecessary since there will be no more houses in low-lying areas (Suppasri et al, 2015). Disagreements regarding the height of seawalls remain, as well as environmental concerns regarding the unprecedented amount of debris and soil moved from the leveled mountains to raise the downtown areas of some coastal towns (Aldrich, 2019).

Fourthly, many are worried about tax pressure on residents in the years to come. In 2013, the Japanese government levied a 2.1% income tax surcharge nationwide to cover post-tsunami recovery, slated to last 25 years (Pinter et al, 2019). On top of this tax, residents of coastal Tohoku expect their own taxes to increase. Indeed, while the construction of the sea defenses falls on the government and prefectures, maintenance will be covered by residents.

A fifth issue concerns the effects of walls on mortality. In his book *Black Wave*, Aldrich (2019) analyzed mortality statistics related to the disaster, concluding that seawalls offered little to no protection against the tsunami. Independent reports stated that in some cases, sea defenses may have amplified the power of the waves and caused more damage to communities (Onishi, 2011). While hard structures contributed very little, it was social infrastructure—the social pressure to evacuate—that saved lives (Aldrich, 2019).

A sixth problem relates to environmental concerns. The construction of sea defenses in Japan is not subject to the national Environmental Impact Assessment Law, further exacerbating discontent and concern for the environment. Such projects will impact the physical environment along the coast forever. Yet, as Aldrich (2019) noted, 'Regulations allow developers and construction firms to move these projects forward without even paper studies of their potential influence on ecosystems' (p. 140).

Lastly, and perhaps most importantly, residents did not feel that they were involved in the design of the new sea defenses. They felt invisible and that they had no voice or choice in the matter. Abe-san, a resident of Otsuchi interviewed by the author, explained that meetings organized by the government to explain the megaproject were systematically scheduled at times when most of the adult population was at work. He said, 'There was no voting process, they didn't really listen to us. People were worrying. (...) But since only the elderly were present, that group (...) they generally agree with whatever the government says.'

Destroying the continuity between land and sea would be destructive to the social, religious, and economic lives of these communities. Many people have been asking for designs that are not concrete-based, but evacuation-based, rediscovering alternative tools and practices to manage hazardous situations. While the construction of seawalls creates a separation between residents and the sea, the proposed alternatives are seen as more consistent with historical and religious traditions: The water is not perceived as an enemy from which the towns must be protected and separated, but a neighbor that is sometimes friendly and sometimes aggressive. However, the sea is never something to get rid of since the livelihood of these fishing communities depends on their relationship to it.

Figure 4.1 | Naoya Takahashi in front of one of the new seawalls. Photograph by Camilla Martini.

Figure 4.2 | Seawall in Kesennuma, Miyagi, next to the fish market. Photograph by Camilla Martini.

Where is Hope? (Co)Designing for Better Futures

To address the unpredictable nature and increasing size of catastrophes today, seawalls should be designed with the assumption that they can be destroyed and that overtop will occur. The core discourse among affected residents, gathered from several fieldwork trips in the area, is that the government's standardized design did not account for the fact that most were willing to accept hazards as part of their lives (since it is indeed part of the cultural history of a country in which tsunamis and earthquakes are a regular occurrence). When resistance to the seawalls was met with the metaphorical wall of Japanese bureaucracy and rebuffed, some communities gave up. However, others kept fighting for both shorter seawalls and the addition of alternative practices and tools for tsunamis that they felt maintained a more harmonious relationship with the landscape and their traditions. These traditions included their jobs as fishermen, using the ocean as an integral part of rituals and religious festivals in their communities, and sea-based tourism.

In Minamisanriku, residents used to meet at the Kaminoyama Hachimangu Shrine to discuss the reconstruction and how to avoid—or at least modify—the sea defense project for the municipality. Mayumi Kudo, the temple's *Shinto* priestess, said, 'We noticed that at one of the beaches in Matsubara, the marine fauna—and in particular some baby oysters—were coming back. The plan was to build the seawall right in front of that area. For us, it would have been better not to have a barrier at all. But since we were in no position to make demands, we only asked for that spot to be preserved' (personal interview, 2020). After much discussion and negotiations, the request was accepted and the position of the seawall was modified to encircle Matsubara Beach instead of cutting through it and making the environment unlivable for the oyster colony. Negotiating a design mindful of non-human marine elements is crucial to maintaining both the jobs of fishermen and the town's traditions. When designed effectively, the seawall itself can become a vehicle for the settlement and maintenance of local marine ecosystems. For example, some new ecological designs demonstrate how smooth vertical concrete coastal infrastructure, such as many types of seawalls or breakwaters in the marine environment, have no positive ecological value and should only be considered a last resort, with rough substrates being preferable due to their ability to facilitate the settlement of organisms (Piocha et al, 2018). As part of disaster risk reduction and adaptation strategy, ecological engineering and ecosystem-based approaches hold a lot of potential for the design of sea defenses in terms of both costs and potential regret related to the impacts of built structures on the landscape and environment (Furuta and Shimatani, 2018).

Mayumi Kudo is also active in preserving the collective memory of 2011 as well as of previous disasters. She has written books and created a scroll that lists every major disastrous event that has occurred in the area over the past thousand years, showing the amount of damage and how people recovered. To maintain the sense of danger in decades when tsunamis and unpredictable events are absent—in addition to physical structures such as museums and memorial stones (which historically have helped increase tsunami awareness and reminded people to evacuate to safe areas)—some of the towns have *kataribe*, traditional storytellers that provide testimonies on the disaster and educate tourists on the 2011 disaster. These practices also include a 'heterogeneous collection of local sayings, behavior patterns, and social arrangements, called *saigai-bunka* (disaster culture)' (Kimura, 2016, p. 28). For example, in Rikuzentakata, the area inundated by the tsunami has become a giant memory park in which the sleek Iwate Tsunami Memorial Museum is integrated with a promenade that leads up the new seawall.

Tomoyuki Miura, elected to the council of the Oyakaigan (Oya Beach) district of Kesennuma, also echoed the general sentiment: 'Many of us believe that the wall is not necessary,' Miura explains. 'The best solution would be to think of alternatives such as the development of more efficient escape routes, and favoring the construction of houses in the more elevated areas. The fact is, the reconstruction plan establishes that the money has to be allotted firstly for the wall, and only thereafter for any urban planning project' (personal interview, 2020). The reconstruction plan proposed by the government dictates that the possibility of utilizing funds for urban reconstruction—i.e., of roads and houses—is tied to the acceptance of the megaproject. The residents of Oya Beach wanted to preserve their shoreline, which would have been cut in two by a massive 9.8-m-high seawall. Local residents asked for the wall to be moved back from the beach. As Pinter et al (2019) explained, 'Objections to this modest request did not focus on engineering challenges or cost, but instead were jurisdictional—seawalls in Japan are under the control of the Forestry Agency, whereas national highways are managed by the Ministry of Land, Infrastructure, Transport, and Tourism. No shared project was considered possible' (p. 1434). In the end, a combination of media attention to the issue and lobbying by local stakeholders allowed the wall to be built further inland and preserve the shore—creating a hybrid and multifunctional design in which a highway was built on top of the seawall, thus using the seawall funds partly for urban reconstruction.

The town of Iwanuma also adopted a multifunctional approach by utilizing the debris of the tsunami for a project called Millennium Hope Hills. The debris was used to shape 15 hills approximately eight m high, on which trees have been planted. While there are short seawalls on the coast, the real embankment project is that of the hills that overlook the city, giving a sense of being surrounded by nature instead of a stark barrier. The hills also host a memorial and recreational park, and the project has garnered strong support from residents and the media (Suppasri et al, 2015).

Figure 4.3 | A detail in the Kesennuma seawall: A hole to look at the sea. Photograph by Camilla Martini.

(Co)Designing with natural elements, which are less invasive and create continuity with the landscape, has been partly integrated into official designs, mostly in the form of the so-called forest walls, which maintain a softer barrier between communities and the sea. Coastal forests act as buffer areas against tsunamis, increasing resistance and giving residents more time to evacuate (Mori, Yoneyama and Pringle, 2015). Planting forests behind seawalls (the most common option in Japan) is considered a long-term project since they need decades to grow and set their roots in the ground so that they can withstand or minimize the impact of a tsunami. This constitutes a return to the past, in which the Japanese coastline was lined with pine forests that naturally created a protective barrier. Additionally, during non-tsunami periods, forest walls consisting of black pine trees are considered a nature-based mitigation method used to collect blown sand, mitigate wind speeds, and protect farmlands and residential areas. Most cities had to negotiate design co-creation in a situation in which the disruptive element—the walls—could not be removed, and they did so not only to mitigate the impact of these forms of concrete infrastructure but also as a way of re-appropriating the landscape and fostering hope. Towns like Rikuzentakata have planted forests between the sea and the wall. This is most likely because the seawall is incorporated into the built system that includes the Iwate Tsunami Memorial Museum. The seawall can be reached and walked onto to look at the other side and enjoy the view. Thus, it seemed more appealing to plant the forest beyond the wall and toward the ocean.

While the government has planned for uniform seawalls in every affected community (with slightly varying heights), other voices have suggested that what is needed are non-structural designs that consider the peculiarities of each town (Kimura, 2016) and its human and non-human elements. Moreover, engineering studies have confirmed that a hybrid infrastructure accounting for both a built part (the wall) and a natural part (the forest) could be the best way to protect against unpredictable hazards (Omori et al, 2022).

While the 'issue with walls' seems to be a quantitative issue of height at the surface level (Kimura, 2016), it is actually a qualitative question of horizontal multivocality expressed by establishing a set of disaster-related practices and tools that can not only block a tsunami of expected height but also foster rapid evacuation while minimizing damage from unpredictable disasters, even when the wall is overtopped or destroyed. Fishermen's knowledge and practices are inseparable from the landscape in which they are embedded. Thus, while creating a barrier might block seawater from coming in, it will most likely accelerate the hemorrhage of (especially young) people leaving these communities for a better life in the Japanese metropolises. What is at stake with the seawall megaproject and the residents' reactions to it is a clash between technocentric designs that rely on the past and socio-ecologically

Figure 4.4 | View from the Rikuzentakata seawall. A forest wall has been planted in the area in front of the seawall. Photograph by Camilla Martini.

centered designs that give agency back to the communities, implying a certain degree of presumed vulnerability that enhances the ability and responsibility of citizens to evacuate in case of disaster and simultaneously maintains a relationship with their landscape in daily life. Matanle, Littler and Slay (2019) invoked the Japanese environmentalist Tanaka Shōzō, who urged Japanese people to adapt to the flow of nature rather than defend against it, lest they are undone by the force of its backflow. Sea defenses follow an overconfident pre-2011 rationale that fails to consider the changes and adaptations necessary today.

Acknowledgments
This project has been developed and realized thanks to the funding offered by the Landscape Research Group and the collaboration of Camilla Martini and Marco De Vidi, who have helped by conducting interviews and collecting pictures and videos.

References

Aldrich, D.P. (2017) 'Trust deficit: Japanese communities and the challenge of rebuilding Tohoku.' *The Asia Pacific Journal: Japan Forum*, 29 (1), pp. 39–52.

Aldrich, D.P. (2019) *Black Wave*. Chicago, University of Chicago Press.

Furuta, N. and Shimatani, Y. (2018) 'Integrating ecological perspectives into engineering practices: perspectives and lessons from Japan.' *International Journal of Disaster Risk Reduction*, 32, 87–94.

Kimura, S. (2016) 'When a seawall is visible: Infrastructure and obstruction in post-tsunami reconstruction in Japan.' *Science as Culture*, 25 (1), pp. 23–43.

Matanle, P., Littler, J. and Slay, O. (2019) 'Imagining disasters in the era of climate change: is Japan's seawall a new Maginot Line?' *The Asia-Pacific Journal: Japan Focus*, 17 (13), pp. 1–29.

Ministry of the Environment of Japan (2001) *5th Basic Survey on Conservation of Natural Environment*. Available at: <http://www.biodic.go.jp/kiso/99/umibe.html/> [Accessed 10 September 2022].

Mori, N., Yoneyama, N. and Pringle, W. (2015) 'Effects of the offshore barrier against the 2011 off the Pacific coast of Tohoku earthquake tsunami and lessons learned.' In: Santiago-Fandiño, V., Kontar, Y.A. and Kaneda, Y. eds. *Post-Tsunami Hazard*. New York, Springer, pp. 121–132.

Omori, Y., Kuriyama, K., Tsuge, T., Onuma, A. and Shoji, Y. (2022) 'Coastal community preferences of gray, green, and hybrid infrastructure against tsunamis: a case study of Japan.' In: Nakamura, F. ed. *Green Infrastructure and Climate Change Adaptation*. Singapore, Springer, pp. 415–441.

Onishi, N. (2011) 'Seawalls offered little protection against tsunami's crushing waves.' *New York Times*. Available at: <https://www.nytimes.com/2011/03/14/world/asia/14seawalls.html> [Accessed 25 September 2022].

Pinter, N., Ishiwateri, M., Nonoguchi, A., Tanaka, Y., Casagrande, D., Durden, S. and Rees, J. (2019) 'Large-scale managed retreat and structural protection following the 2011 Japan tsunami.' *Natural Hazards*, 96 (3), pp. 1429–1436.

Pioch, S., Relini, G., Souche, J. C., Stive, M.J.F., De Monbrison, D., Nassif, S., Simard, F., Allemand, D., Saussol, P., Spieler, R. and Kilfoyle, K. (2018) 'Enhancing eco-engineering of coastal infrastructure with eco-design: Moving from mitigation to integration.' *Ecological Engineering*, 120, pp. 574–584.

Raby, A., Macabuag, J., Pomonis, A., Wilkinson, S. and Rossetto, T. (2015) 'Implications of the 2011 great East Japan tsunami on sea defence design.' *International Journal of Disaster Risk Reduction*, 14, pp. 332–346.

5 | Imaginaries of Resilience

Rhetorics and Realities of Climate Change Adaptation in Four Southeast Asian Megacities

Lizzie Yarina

Design Project Data

1 | **Bangkok City**
Location
Bangkok | Thailand | Asia
Coordinates
13° 45' 14.33" N | 100° 30' 5.18" E

2 | **Ho Chi Minh City**
Location
Ho Chi Minh City | Vietnam | Asia
Coordinates
10° 49' 22.87" N | 106° 37' 46.74" E

3 | **Jakarta City**
Location
Jakarta | Indonesia | Asia
Coordinates
6° 12' 0.00" S | 106° 48' 59.99" E

4 | **Manila City**
Location
Manila | Philippines | Asia
Coordinates
14° 35' 58.2" N | 120° 59' 3.12" E

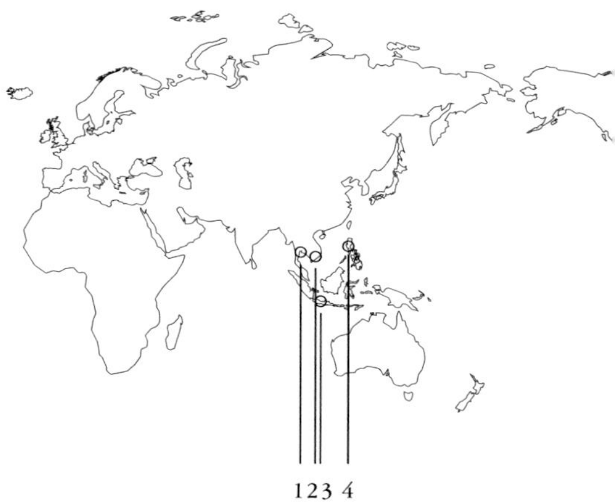

Across Southeast Asia, coastal and deltaic megacities are threatened by the confluence of rapid growth, rising seas, and subsiding ground levels. As local governments seek to adapt their cities to these broad risks, foreign experts have offered their services, bringing *climate-proofing* models from their respective countries. But whose resilience is really being promoted by these plans? How does the positivist rhetoric of climate change adaptation obscure damages to the most vulnerable citizens? This chapter investigates four of the largest and most climate-vulnerable cities in Southeast Asia: Ho Chi Minh City, Manila, Jakarta, and Bangkok. All have populations exceeding eight million, occupy low-lying deltas, and suffer from repeated inundation. Although they have vast differences in their urban structures and socio-political contexts, each is developing a set of responses to escalating flooding, with such plans being implicitly or explicitly tied to climate change. Typically, these plans involve collaborations with foreign firms or bi-/multilateral organizations. Building on this volume's emphasis on the roles of hope and imagination, the present chapter also examines how flood-exposed communities across these sites have imagined alternative modes of adaptation that emphasize the capacities of communities to imagine their own climate futures. What is the potential of these alternative imaginaries of resilience?

Introduction

In 2011, a catastrophic flood washed through greater Bangkok. From a hydrological perspective, this was not an unusual event since Bangkok occupies the Chao Phraya River Delta. Although the rainfall that year was higher than normal, the waters did not reach the hundred-year flood level. However, the landscape was more vulnerable than in past cycles (Ziegler et al, 2012). Factory development in the floodplain, subsidence caused by groundwater extraction, and the mismanagement of dams upriver led to severe flooding that killed more than 800 people and affected some 13 million lives. Protected by the King's Dike, which encircles the Bangkok Metropolitan Area, the city itself was largely spared; however, displaced floodwaters made conditions worse in outlying neighborhoods and upstream cities. The sacrificial zones were inundated for weeks, and then months. As angry 'flood mobs' from the outer districts descended on the flood barriers, opening sluice gates and tearing holes in the sandbag walls, the prime minister counseled them to think of the national good. If the city center flooded, she said, it would cause 'foreigners to lose confidence in us and wonder why we cannot save our own capital' (Winn, 2011).

Herein lies a dark truth about planning for *climate resilience.* Decisions regarding which areas will be protected are not only about whose safety will be guaranteed; they also involve transnational concerns such as reassuring global investors and preserving manufacturing supply chains. In Thailand, thousands of soldiers were dispatched to patrol the floodwalls. They were *enforcing* resilience. This is both a rational decision and a disturbing vision of our climate-altered future. We are heading toward a world in which the unequal distribution of environmental risks is administered by state violence. How did we get here?

This article investigates the cases of four large cities in Southeast Asia facing major climate risks: Jakarta, Manila, Ho Chi Minh City, and Bangkok. Each city is home to at least eight million people living in a low-lying delta threatened by rapid urbanization, sinking ground, and rising seas. As officials seek to make their cities more resilient, they bring in external planning experts who push *climate-proofing* models developed in Japan and Europe. Many such European experts are from the Netherlands, which has a long history of advanced water management strategies. Highly engineered, technocratic programs come with readymade slogans, such as 'making room for the river,' a concept that works well along the banks of the Rhine but can mean mass evictions in the Global South. When 'slums' (often a slur for urbanized villages with deep histories) are represented as a blight to be scraped away with little if any recompense and their people resettled in untenable locations far from the city center, we must ask the following question: 'Whose resilience' is really being promoted (Vale, 2014)? Too often, the rhetoric of climate adaptation is doublespeak for the displacement of poor, informal communities, as well as an alibi for unsustainable growth.

Despite these issues, planning for climate change remains critical. We should strive to build cities and landscapes that are socially and ecologically healthy, and the interventions of planners and designers can help us adapt to uncertain future conditions. Flood-exposed communities across these sites have imagined alternative modes of adaptation that emphasize their capacity to imagine new climate-altered futures from the ground up. Alongside examining the limitations of top-down adaptation planning, this chapter closes by considering locally-driven imaginaries of resilience and the potential of recombining strategic and tactical approaches in the production of resilient futures.

Jakarta

Let us begin with Jakarta, where the Great Garuda is the charismatic megafauna of resilience infrastructure. Approximately 40 percent of the city is below sea level, and regular flooding along the highly polluted rivers and colonial canals is a fact of life. In 2007, a particularly heavy flood forced 300,000 evacuations and spurred new plans to fortify the city against rising waters. An international team led by the Dutch engineering firms Witteveen+Bos and Grontmij proposed building the National Capital Integrated Coastal Development (NCICD), which envisioned artificial islands in Jakarta Bay anchored by the world's largest sea wall. The scheme, which resembles a garuda—a mythical bird and the national symbol of Indonesia—was to largely be financed by private development on the islands, including a new central business district housing 1.5 million people.

The project's manager described the NCICD as 'one big polder,' referencing the Dutch strategy for enclosing land within dikes to artificially control its hydrology (Interview, 2017). Essentially, Jakarta Bay would be a bathtub completely separated from the Java Sea; the city's rivers would drain here and then be pumped out to the ocean. Critics argued that disrupting the hydrology would harm local fisheries, trap polluted waters within the city, and exacerbate flooding outside the wall. In response to these concerns, as well as allegations of corruption, the project has since been scaled down. If the bay is polderized, it will be Jakarta's iron lung, requiring a whole secondary life-support system of pumps and drainage systems. To prevent retention areas from becoming polluted 'black lagoons,' the city will require major sanitation upgrades, which are being led by German and Japanese partners. Getting the various projects to function well together on time and within scope is an immense challenge.

One of these parallel works is a 'room for the river' plan known as the Jakarta Urgent Flood Mitigation Project, approved in 2012 with support from the World Bank. In the Netherlands, making room for the river means clearing land in the floodplain and moving back dikes to accommodate the river's fluctuations. In Jakarta, the government must first evict some 200,000 people from *kampungs*, which are urban villages along the river where land tenure rights are unclear. As places with histories of their own, kampungs have expanded deeper into urban waterways, where they hinder flood management. Sandyawan Sumardi of the advocacy group Ciliwung Merdeka explains that many residents are caught in an ongoing cycle of evictions, moving to a new kampung only to see it cleared again months later (Interview in Transformasi Indonesia, 2014). City officials refer to these evictions by the Orwellian term 'river normalization,' and they are occurring more frequently as the sea wall project advances. Notably, effective drainage of the city's rivers will be essential to the operation of the polder system (Goh, 2021).

Many kampungs along the rivers are now piles of brick and rubble. Residents who have not yet been allotted resettlement housing, or who have determined that the off-city sites were unlivable, attempt to rebuild homes and shops atop the debris of their former lives. Pluit reservoir, one of the retention and pumping areas in the NCICD master plan, is already a 'black lagoon.' Debris from upstream accumulates here, and the reservoir resembles a wet landfill overrun by water hyacinth. Backhoes precariously balanced on makeshift barges chip away at detritus to keep the water flowing, while garbage trucks dump more waste directly into the lagoon.

We often perceive environmental risk in coastal cities through the lens of climate change. If seas are rising, a great sea wall will hold them back. However, in Jakarta, the ground is sinking faster than the sea is rising—the delta is urbanizing rapidly, and the weight of construction compacts the alluvial soil while population growth leads to the over-extraction of groundwater. This subsidence can clearly be seen in the Pluit neighborhood. Walking on top of the sea wall puts you at eye level with boats in the harbor, while on the other side, kampung life bustles a full story below. Needless to say, the neighborhood floods regularly with heavy rainfall and king tides. While subsidence has repeatedly been documented as a major risk factor, officials have taken little action to address it (Deltares, 2013). Notably, projects that can be sold as climate adaptation seem to have priority over efforts to mitigate groundwater extraction (Bakker, 2007).

Manila

The World Bank is also funding slum clearance projects in Manila, an extremely dense coastal city with 24 million people in its metropolitan area. Although Manila is sinking by nearly five cm per year (Deltares, 2013), subsidence here is a lesser concern than storm surges and typhoons. Heavy rain causes flooding throughout the city, particularly as floodplains and

Figure 5.1 | Coastal sea wall in Kampong Pluit, Jakarta. Photograph by Lizzie Yarina.

distributaries are paved over for development. Manila is already battered by several typhoons each year, and the storms are projected to grow stronger and more unpredictable with climate change. In 2013, Super Typhoon Yolanda, one of the most powerful storms in recorded history, killed more than 6,000 people in the Philippines.

The Metro Manila Flood Management Master Plan, approved in 2012, involves physical strategies such as dams, dikes, canal-dredging, new pumping stations, and the 'resettlement of informal settlers that have encroached on the waterways' (World Bank, 2017). While not explicitly a climate adaptation plan, one of its proposed outcomes is 'increased resilience to natural disaster and climate change impacts.' The US $8 billion master plan is led by a Dutch irrigation engineer, and the World Bank describes the project as 'learning from the experience of the Netherlands' through 'a new approach that *makes room for the river to flood*' (World Bank, 2014). Although a senate report describes this as a 'modern' and 'ecological' approach to managing watersheds, the report also defines the clearance zone in terms of fixed distances based on the 1976 Philippine Water Code rather than ecological parameters (Senate of the Philippines, 2014).

This is a dramatic redefinition of the city's relationship with its rivers. Historically, people in Manila have lived in flood zones but managed risk through adaptive architectures, such as the traditional *bahay kubo* house, which is made of light materials and raised on stilts.

The plan targets 125,000 households for eviction, almost exclusively in the so-called slums along waterways, even though there are also wealthy neighborhoods and new developments in the floodplains (Anguelovski et al, 2016). 'Operation Evacuation' initially emphasized voluntary relocation to sites within the city where residents could maintain existing social and economic ties; however, with affordable urban land now scarce, the program involves forced eviction ('involuntary resettlement,' in World Bank parlance) to resettlement sites that are two to four hours away (Mabilin, 2014).

When residents resist being moved, evictions can turn violent, as in Sitio San Roque, where 1,000 armed officers descended on a barricaded neighborhood and demolished it. Although the displacement is framed as being for the residents' own good, Al Bernarte of the advocacy group Urban Poor Associates compared the resettlement sites to refugee camps and noted that they disrupt the social networks and economies people have built their lives around. Sometimes the new housing is no safer than the old. At Kasiglahan Village, a resettlement site located in a flood plain and known geohazard zone, nearly 2,000 homes were flooded in a storm (Advocates of Science and Technology for the People, 2014). Bernarte also stated the following: 'We are not getting people away from danger and putting them in a safer place. We are just transferring them from one risky area to another' (Interview, 2017). Cut off from jobs, friends, and families, half of the evictees abandon their resettlement housing and return to the city (Department of Public Works and Highways and Metro Manila Development, 2014; Alvarez and Cardenas, 2019).

Ho Chi Minh City

Ho Chi Minh City (also known as Saigon) sits on the edge of the Mekong River Delta, a 40,000-square-kilometer zone made ecologically and agriculturally rich by the fertile soils carried by distributaries that fan out toward the sea. Humans in the delta have long regulated water levels using small canals and dikes. In addition to land-based agriculture, floating rice crops and fisheries are also utilized (Biggs et al, 2009). During the French colonial era, a more comprehensive strategy of 'Dutch dikes' and polders was developed, which was further intensified during the 'rice everywhere' campaigns of the 1970s and 80s. While flooding is part of rural systems, it brings life to a standstill in the city. Frequent floods leave homes waterlogged and commuters pushing motorbikes through waist-deep water. Dr. Ho Long Phi, director of the Center for Water Management and Climate Change at Vietnam National University, has observed that over the past decade, the city has begun to experience a new sort of flooding, 'rising year by year' in terms of its frequency, duration, and extent (Interview, 2017).

The city is slated to spend US $4.4 billion on flood reduction measures, including improvements to drainage systems and sewers, tidal control projects, dikes, and reservoirs (Global Construction Review, 2016). Unsurprisingly, many of the projects are driven by Dutch partners, who have helped draft the *Ho Chi Minh City (HCMC) Climate Adaptation Strategy*, which strains to compare the Vietnamese city with its adopted cousin, Rotterdam: 'Both cities are located in the delta of a large river system, have a large port and are an important factor in national GDP' (Vietnam Climate Adaptation Partnership, 2013). The focus on these superficial similarities belies vast differences in social, economic, and institutional structures. The document conveniently ignores Ho Chi Minh City's large informal population, its lack of political freedom, and the complexities of land tenure in a nation where private property is an emerging concept—except, of course, when it blames slums for encroaching on waterways.

The original plan envisioned a US $2.6 billion ring dike that would stretch 172 km, wrapping the urban area with flood protection infrastructure moderated by sluice gates that would allow canals to empty into the Saigon River, along with 12 large tidal barriers. In response to criticism that the proposal was intentionally over-scaled to boost kickbacks for bureaucrats, two Dutch firms were commissioned to develop a cheaper and slightly smaller version. Public opposition and a lack of funding have prevented either proposal from being implemented in full, but some components are already under construction, such as an eight km dike along the Saigon River set up through a public-private partnership with a Vietnamese developer who will receive land rights in exchange (Thanh Nien News, 2015). More than 1,500 households will be evicted, and the project has been criticized for increasing flood risk for the more rural (and poorer) neighborhoods on the 'wrong' side of the wall (Scussolini et al, 2017; Anh, 2016).

As with plans in other cities, the *HCMC Climate Adaptation Strategy* is filled with ideas that may make sense in a Western context but fail to translate smoothly to local conditions. For example, it states that dikes should 'be located leaving as much for the river and its flood plains as possible' (Vietnam Climate Adaptation Partnership, 2013), which makes sense until you realize that what is 'possible' here depends on whether the government recognizes the land rights of thousands of people. The pilot plan for District Four gives the district its own small ring dike that neatly eviscerates the island's perimeter of informal housing while allowing the sale of land for new development atop and around the dikes. The strategy document states that this plan was created in collaboration with the community, but a white paper clarifies that workshops were held with government staff and 'additional experts' as stakeholders (Lasage et al, 2014). Since there is no right to free speech in Vietnam, true community participation is unlikely. The displaced are offered settlement sums that are too low to buy housing near their former communities, or possibly anywhere in the city.

Bangkok

This brings us back to Bangkok, where there has been little progress toward a comprehensive plan for climate adaptation since the catastrophic floods of 2011. The immediate response was not to build more resilient housing but to harden the defensive infrastructure in industrial areas to discourage investors from fleeing the country. Bangkok is constantly fortifying against flood risk created by its very existence, thus exemplifying Ulrich Beck's (1992) framing of a 'risk society' in which modern humans are caught in a cycle of mitigating the risks that they themselves have created. In decades past (and still today in some rural parts of the country), Thai people lived in 'amphibious communities;' for example, in 'raft houses' that float upwards on stilts during floods or in villages built on two levels, where upper walkways and living quarters can be used during the rainy season (Jumsai, 2011). However, such adaptive patterns are disappearing, even as the country faces greater climate risks.

Apiwat Ratanawaraha, a lecturer in planning at Chulalongkorn University, says the military government (like its predecessors) sees flooding as a 'water infrastructure issue,' with little regard for the complex layering of social and spatial factors (Interview, 2016). In the 1980s, Dutch and Japanese engineers helped build a polder system protected by a 72-km dike. However, as the city sprawled, urbanizing districts outside the polder faced more severe flooding than if there had been no dike at all. After heavy floods in 2006, engineers drew a bigger circle expanding the King's dike to the boundary of the Bangkok Metropolitan Authority and raising its height in some areas, which again intensified problems for those outside the line in 2011. There are also floodwalls along 80 km of the Chao Phraya River and a system of pumps that help mediate the stormwater and urban water network—so long as the power stays on. While the official policy is to 'keep Bangkok dry at any cost' (Lebel et al, 2012, p. 283), this is an impossible task given the city's hydrology.

Efforts are ongoing to reclaim a large swathe of riverfront land for the ironically named 'Chao Phraya for All' megaproject. Featuring a transportation corridor and riverwalk green spaces, the master-planned development will stretch 57 km along the river behind a 3.5-m-high levee (the first phase, involving 14 km, was slated to open in late 2018). As in other cities, this will involve clearing 'illegal' housing in dozens of communities along the river. The project has met vocal public resistance, even under Thailand's repressive government. However, the junta has already removed more than half of the homes in the first phase while providing few details about its plans and avoiding engagement with local communities.

Designers opposed to the project say it will disconnect people from the river while exacerbating flood risk on both sides of the wall (Personal communication, Friends of the River). They argue that the high dike will narrow the river channel, accelerate the flow, and raise water levels while also blocking runoff from the highlands, potentially turning 'the inland into a bowl of soup' (Activists with the Chao Phraya Riverside Area Development Alliance, quoted in Kongrut, 2015). Some suggest that the megaproject is not really about building resilience at all, but is rather a veil for corrupt developers and government officials to line their pockets (Elinoff, 2017).

Imaginaries of Resilience

As these four scenarios show, the main problem with the now-ubiquitous discourse on 'resilient cities' is that it assumes a shared understanding of what resilience is. Instead, we should speak of *imaginaries of resilience* or the diverse sets of ideas and practices around adapting to environmental risk. Notably, these imaginaries reflect political and landscape values. Sometimes, the clash of imaginaries is suddenly made visible, such as when flood mobs descended on the King's dike in Bangkok, revealing the tension between the prime minister's vision of a dry urban center projected outward to global media and the residents' insistence that their own homes and neighborhoods be protected. However, such occasions are rare. More often, the competing

Figure 5.2 | Traditional canal houses in Bangkok, alongside the public ferry. Photograph by Evan Krause.

imaginaries of resilience are obscured by the depoliticized language of bureaucratic reports and climate adaptation projects.

Therefore, what are the alternatives to these top-down, technocratic, socio-spatially uneven plans? Where are the projects that elevate local knowledge or return justice to the displaced? In her book *Form and Flow: The Spatial Politics of Urban Resilience and Climate Justice*, architecture and planning scholar Kian Goh uses the term 'counterplan' to describe projects that imagine critical, locally driven alternatives to top-down planning processes (Goh, 2021, p. 157). In Jakarta, kampung-dwellers who were excluded from consultation on the government's normalization and resettlement plans collaborated with architects and universities to promote alternative visions for kampung futures. In this context, Goh argued that 'design becomes part of a platform for organizing' (Goh, 2015, p. 250), a way for communities to visualize and formalize different ways of understanding and constructing the city. These alternatives are needed because megaprojects at the scale of the Great Garuda negate the possibility for discourse. As geographer Erik Swyngedouw observed, 'Any policy intervention, when becoming concretely geographical or ecological, is of necessity a violent act of foreclosure of the democratic political (at least temporarily), of taking one option rather than another, of producing one sort of environment' (Swyngedouw, 2011, p. 273). While megaprojects present a singular solution to the problem of resilience, counterplans emphasize that climate adaptation is a negotiated space.

Urban Poor Associates (UPA) maintains that communities in Manila's 'danger zones' should be able to determine for themselves whether and how they relocate. UPA sees this as a fundamental right. According to Bernarte, people along urban waterways should be informed about the risks and offered appropriate and adequate resettlement assistance; however, if they still choose to stay, he argues for upgrading their homes *in situ*. The organization's ethos is to 'go where the people want to go.' With UPA's support, the informal community at Estero de San Miguel created a People's Plan for upgrading their urban village to be more resilient to floods and earthquakes. They brought their proposal to the national government (taking advantage of their electoral power during an election year) and received approval to build a colorful, energy-efficient, three-story housing complex for 600 people, with floodable commercial space on the first floor (Almendral, 2015).

Stories such as this support UPA's strategy of promoting urban resilience by building tightly-knit community networks in the informal neighborhoods where residents spend the majority of their time. When neighbors form close relationships, they are attuned to cyclical and seasonal shifts in the environment, and also to the changing needs of others. In 2009, when Typhoon Ondoy flooded parts of Manila with up to three meters of water, the informal communities where UPA works had no casualties, while death tolls were highest in wealthier suburban neighborhoods. The Philippines gets hit by as many as 20 typhoons per year, and communities along *esteros* (waterways) often adapt without the help of outside designers by raising floor levels, building bridges, establishing community evacuation centers, and commuting by boat during periods of high water levels (Porio, 2011).

As an alternative to dikes and floodwalls, some architects have proposed climate adaptation models that draw on local vernaculars for living with water. Terdsak Tachakitkachorn, a professor of architecture at Chulalongkorn University in Bangkok, told me that villagers in Ayutthaya incrementally raise their houses 50 cm at a time, adapting the ground plane to peak flood levels (Interview, 2016). In another example, rice farmers in greater Bangkok, who are accustomed to fluctuations in water level, will switch from farming to fishing in times of flood. This form of adaptation is known as 'waterscape urbanism' (Thaitakoo and McGrath, 2010, pp. 35–50). These approaches are significant not only in terms of how they build on local knowledge and priorities but also in how they imagine a future integrated into a changing landscape, rather than holding back water through dikes and levees.

In the allied disciplines of design, architecture, and planning, there has been a lack of critical attention to the discourses of climate adaptation and resilience. Dutch models are simply accepted as best practices or 'portable solutions' that can be applied to any context worldwide. We need more 'counterplans' that give voice to the vulnerable, elevate local knowledge, and reveal alternative models of resilience. This is not to suggest that 'big plans' to mediate environmental risk are never appropriate—inevitably, climate change will require radical conceptions of how we make and re-make our cities. These can present important opportunities to establish new values in terms of how we relate to our environment. However, their conceptualization cannot be left *only* to leaders and experts. Climate change adaptation is a negotiated terrain, with competing imaginaries of resilience rather than a technical solution that leaves no room for contestation. Designers and planners should critically reflect on language suggesting that adaptation projects can be win-win scenarios for both cities and developers and recognize that the emerging financialization of adaptation risks prioritizes growth over collective security. It elides the difference between the 'little d' development of speculative, for-profit real estate construction and the 'big D' development of modernizing 'underdeveloped' societies or spaces. In societies where the forces of global capital have outmaneuvered local politics, megaprojects built under the banner of climate adaptation can overrun those most vulnerable to (and least responsible for) climate change. When planners and designers fall in line with conventional climate-proofing and pro-development narratives, they help to erase the communities that are not their direct clients. Technical experts are primed to apply their training to design problems within their scope and ignore everything else.

Imagining more just approaches to adaptation calls for bringing together strategic visions with tactical, grassroots approaches. For designers, this suggests moving away from the production of grand visions and towards a role of mediating and facilitating design processes (Yarina, 2019; Yarina, Mazereeuw and Ovalles, 2019).

Resilience is not a fundamentally technical question; it is social and political. Pre-determined models will not work on this contested ground. Planners and designers must recognize and negotiate the various values, motivations, and 'resilience imaginaries' across the diverse cities in which they are needed. To adapt to the climate crisis, architectures of hope must negotiate these multiple visions of the future.

Acknowledgments

This chapter is an edited version of the article 'Your Sea Wall Won't Save You,' originally published in *Places Journal* in March 2018 (https://doi.org/10.22269/180327). Fieldwork for this project was completed in 2016–17 with partial support from the Massachusetts Institute of Technology Department of Urban Studies and Planning. Thank you to Claudia Bode, Larry Vale, Miho Mazereeuw, and our colleagues at the Urban Risk Lab who provided input on earlier drafts. Thank you to all of those who took the time to speak to me in Bangkok, Jakarta, Manila, and Ho Chi Minh City, without whom this project would not have been possible. Additional thanks go to Josh Wallaert and the Places editors for their support in the making of this article and for their policy of retaining the copyright of the work with the author.

References

Advocates of Science and Technology for the People (2014) *Structural Integrity and Flood Risk in Kasiglahan Village, Montalban, Rizal: A Report.* Available at: <http://www.agham.org/technical-report/structural-integrity-and-flood-risk-kasiglahan-village-montalban-rizal-report> [Accessed 20 January 2023].

Almendral, A. (2015) 'After years of living as squatters, these Filipino neighbors fought for nice, new homes in their community—And won.' *Yes Magazine.* Available at: <http://www.yesmagazine.org/people-power/filipino-neighbors-fought-new-homes-won> [Accessed 20 January 2023].

Alvarez, M.K. and Cardenas, K. (2019) 'Evicting slums, "building back better:" Resiliency revanchism and disaster risk management in Manila.' *International Journal of Urban and Regional Research*, 43 (2), pp. 227–249.

Anguelovski, I., Shi, L., Chu, E., Gallagher, D., Goh, K., Lamb, Z., Reeve, K. and Teicher, H. (2016) 'Equity impacts of urban land use planning for climate adaptation: Critical perspectives from the global north and south.' *Journal of Planning Education and Research*, 36 (3), pp. 333–348.

Anh, L. (2016) 'HCMC starts work on VND10 trillion flood control project.' *The Saigon Times.* Available at: <https://english.thesaigontimes.vn/48213/HCMC-starts-work-on-VND10-trillion-flood-control-project-.html> [Accessed 20 January 2023].

Bakker, K. (2007) 'Trickle down? Private sector participation and the pro-poor water supply debate in Jakarta, Indonesia.' *Geoforum*, 38 (5), pp. 855–868.

Beck, U. (1992) *Risk Society: Towards a New Modernity*, Thousand Oaks, CA, Sage.

Biggs, D., Miller, F., Hoanh, C.T. and Molle, F. (2009) 'The delta machine: water management in the Vietnamese Mekong Delta in historical and contemporary perspectives.' In: Molle, F., Foran, T. and Kakonen, M. eds. *Contested Waterscapes in the Mekong Region.* London, Routledge, pp. 225–248.

Deltares (2013) *Sinking Cities: An Integrated Approach Towards Solutions.* Available at: <https://www.deltares.nl/en/expertise/publicaties/sinking-cities-an-integrated-approach-towards-solutions> [Accessed 20 January 2023].

Department of Public Works and Highways and Metro Manila Development Authority (2014) *Metro Manila Flood Management Project, "Resettlement Policy Framework"* draft. Available at: <https://www.dpwh.gov.ph/dpwh/sites/default/files/MMFM%20-%20RPF_final_May%201%202017.pdf> [Accessed 20 January 2023].

Elinoff, E. (2017) 'Concrete and corruption: Materialising power and politics in the Thai capital.' *City*, 21 (5), pp. 587–596.

Global Construction Review (2016) *Ho Chi Minh City to Spend $4.4bn to Keep Its Head Above Water.* Available at: <https://www.globalconstructionreview.com/ho-chi-minh-city-spend-44bn-keep-i7ts-he7ad-abo7ve/> [Accessed 20 January 2023].

Goh, K. (2021) *Form and Flow: The Spatial Politics of Urban Resilience and Climate Justice*. Cambridge, MA, MIT Press.

Jumsai, S. (2011) 'We lived with floods before; Why can't we do so again?' *The Nation*. Available at: <https://www.nationmultimedia.com/opinion/We-lived-with-floods-before;-why-cant-we-do-so-aga-30168602.html> [Accessed 20 January 2023].

Kongrut, A. (2015) 'Chao Phraya promenade project faces opposition.' *Bangkok Post*. Available at: <https://www.bangkokpost.com/opinion/opinion/590173/chao-phraya-promenade-project-faces-opposition> [Accessed 20 January 2023].

Lasage, R., Veldkamp, T.I.E., De Moel, H., Van, T.C., Phi, H.L., Vellinga, P. and Aerts, J.C.J.H. (2014) 'Assessment of the effectiveness of flood adaptation strategies for HCMC.' *Natural Hazards and Earth System Sciences*, 14 (6), pp. 1441–1457.

Lebel, L., Sinh, B.T., Garden, P., Seng, S. and Van Truc, D. (2012) 'The promise of flood protection: Dikes and dams, drains and diversions.' In: Molle, F., Foran, T. and Kakonen, M. eds. *Contested Waterscapes in the Mekong Region*. London, Routledge, pp. 305–328.

Mabilin, M. (2014) 'Forced evictions, off-city relocation and resistance: Ramifications of neo-liberal policies towards the Philippine urban poor.' In: Moksnes, H. and Melin, M. eds. *Claiming the City: Civil Society Mobilisation by the Urban Poor*. Uppsala, Uppsala Universitet, pp. 133–138.

Porio, E. (2011) 'Vulnerability, adaptation, and resilience to floods and climate change-related risks among marginal, riverine communities in Metro Manila.' *Asian Journal of Social Science*, 39 (4), pp. 425–445.

Scussolini, P., Tran, T.V.T., Koks, E., Diaz-Loaiza, A., Ho, P.L. and Lasage, R. (2017) 'Adaptation to sea level rise: A multidisciplinary analysis for Ho Chi Minh City, Vietnam.' *Water Resources Research*, 53 (12), pp. 10841–10857.

Senate of the Philippines (2014) 'Committee Report No. 73.' *Overall Master Plan on Flood Management in the NCR*. Available at: <http://legacy.senate.gov.ph/lisdata/1972116846!.pdf> [Accessed 20 January 2023].

Swyngedouw, E. (2011) 'Depoliticized environments: The end of nature, climate change and the post-political condition.' *Royal Institute of Philosophy Supplements*, 69, pp. 253–274.

Thaitakoo, D. and McGrath, B. (2010) 'Bangkok liquid perception: waterscape urbanism in the Chao Phraya River Delta and implications to climate change adaptation.' In: Shaw, R. and Thaitakoo, D. eds. *Water Communities Volume 2*. Leeds, Emerald Group Publishing Limited, pp. 35–50.

Transformasi Indonesia (2014) 'The Reality of the Ciliwung People in Jakarta.' *YouTube*. Available at: <https://www.youtube.com/watch?v=fIW3upWHSK8> [Accessed 20 January 2023]. The interview with Sandyawan Sumardi is at 5:35.

Thanh Nien News (2015) *Ho Chi Minh City Assigns Private Investor to Huge Anti-Flooding Project.* Available at: <http://www.thanhniennews.com/society/ho-chi-minh-city-assigns-private-investor-to-huge-antiflooding-project-57179.html> [Accessed 20 January 2023].

Vale, L.J. (2014) 'The politics of resilient cities: whose resilience and whose city?' *Building Research & Information*, 42 (2), pp. 191–201.

Vietnam Climate Adaptation Partnership (2013) *Ho Chi Minh City Climate Adaptation Strategy.* Available at: <http://www.vcaps.org/assets/uploads/files/HCMC_ClimateAdaptationStrategy_webversie.pdf> [Accessed 20 January 2023].

Winn, P. (2011) *In flooded Bangkok, Shades of Hurricane Katrina, The World from PRX.* Available at: <https://www.pri.org/stories/2011-11-01/flooded-bangkok-shades-hurricane-katrina> [Accessed 13 October, 2017].

World Bank (2014) *A Paradigm Shift for Guarding Delta Cities Against Floods.* Available at: <https://www.worldbank.org/en/news/feature/2014/10/08/a-paradigm-shift-for-guarding-delta-cities-against-floods> [Accessed 20 January 2023].

World Bank (2017) *Project Highlights: Metro Manila Flood Management.* Available at: <https://www.worldbank.org/en/country/philippines/brief/project-highlights-metro-manila-flood-management> [Accessed 20 January 2023].

Yarina, L. (2019) 'Critical spatial practice and "resilient" design.' *Society and Space.* Available at: <https://www.societyandspace.org/articles/critical-spatial-practice-and-resilient-design> [Accessed 20 January 2023].

Yarina, L., Mazereeuw, M. and Ovalles, L. (2019) 'A retreat critique: Deliberations on design and ethics in the flood zone.' *Journal of Landscape Architecture*, 14 (3), pp. 8–23.

Ziegler, A.D., Lim, H.S., Tantasarin, C., Jachowski, N.R. and Wasson, R. (2012) 'Floods, false hope, and the future.' *Hydrological Processes*, 26 (11), pp. 1748–1750.

6 | Water Communities in the 21st Century

The Case of Het Lankheet Estate

Inge Bobbink and Amina Chouairi

Design Project Data

Location
Haaksbergen | The Netherlands | Europe
Project Team
Strootman Landschapsarchitecten (Berno Strootman, Rob Aben, Jan Maas, Ingrid Meus, and Thijs van der Zouwen) in collaboration with the Research Center of Wageningen University.
Date
2005 | 2007
Coordinates
52° 13' 72.00" N | 6° 72' 15.16" E
Area
0,06 km^2
Length
n.a.
Investment
n.a.

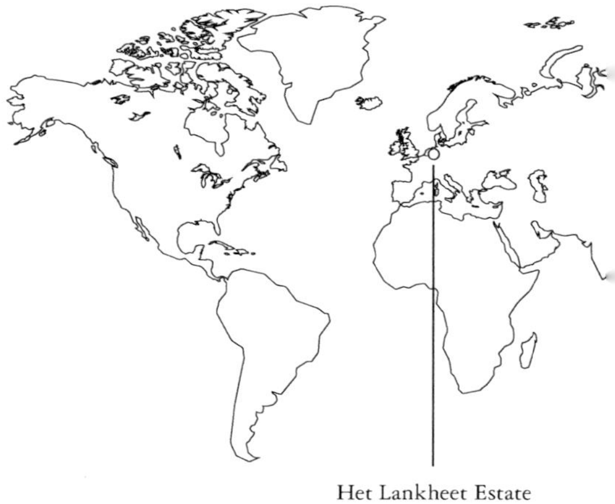

Long-lasting water systems are landscape architectonic projects at their best. The availability of fresh water was often the first motive to settle in certain areas. Through trial and error, communities have developed numerous methods of transforming, expanding, adapting, and reshaping water sources. In particular, when water systems are employed in multifunctional ways, the cultivation practices are in a constant mode of adaptation and monitoring of the system's loading capacity in terms of water amount and quality. Water communities intensely relate to their landscape. After all, the inhabitants' care practices originate from the landscape's sustaining role since the landscape feeds and shelters them. With industrialization and the increasing number of inhabitants, the direct relationship between humans and the landscape in many areas has become lost, thereby impacting the roles of communities in adapting and maintaining water systems.

This chapter explores an initiative in which a newly established community in the Netherlands reactivates an ancient water meadow landscape at the Lankheet Estate, Haaksbergen. The strategy adopted at this estate has assimilated the long-lasting lessons and good practices of communities, becoming a project of hope in the face of today's climate crisis.

Learning From Long-Lasting Water Systems

Human-made water systems are exciting objects of study because knowledge is stored on many levels—technically and socially—offering insights and inspiration to confront today's climate crisis. This chapter focuses on the importance of social involvement relating to the use and adaptivity of water systems, with the insights springing from the Circular Water Stories research conducted at the Delft University of Technology (Circular Water Stories, 2020). As a spin-off of the dissertation *The Landscape Architecture of the Polder-boezem System, Structure and Form of the Water Network, Water Pattern and Water Work in the Dutch Lowlands*, the research shifted from a *solely* spatial-oriented approach to a more holistic interest in global water systems (Bobbink, 2016).

To successfully learn from long-lasting water systems, the illustrative method was developed to study and interpret water systems using drawings and diagrams following guidelines, considering climate conditions, catchment areas, and site-specificity as critical topics to address. During the past five years, a wide range of water landscapes and communities has been analyzed with help of the Illustrative Method by graduate students of the Landscape Architecture Master Track, generating a diverse palette of insightful drawings (Bobbink, Chouairi and Di Nicola, 2022).

Researching water systems worldwide has unveiled how indigenous people use different water types cleverly and circularly. One of the primary inspirations is the wise water usage in Kampung Naga, located in West Java, Indonesia. Spring, rain, and river waters shape the landscape and punctuate Sundanese life. Spring water is collected in cisterns and distributed to water tanks through pipes to serve latrines. The roots of the pristine forest filter rainwater infiltrating the ground. Running continuously, this water is so clean that it can be used as potable water. The constant flow of the river is tapped upstream by opening sluices. There, it is naturally filtered by wild grasses and gravel before being employed for rice fields and fishponds. The community maintains the water system daily, which is an inextricable part of their community's life. For centuries, the Kampung Naga community has secured the system's loading capacity, avoiding potential harm to its functioning and circularity (Tri Prestasia and Kim, 2020).

Long-lasting water systems develop and adapt, reacting to seasonal changes and extreme events such as drought or flooding. People carefully manage these delicate waterscapes as communities live and benefit from them. Moreover, many cultures deeply understand the value of water and worship it.

However, when water systems are no longer maintained, they decline. These systems often suffer due to upgrading techniques involving automatization becoming less contextual and flexible. In the lowlands of the Netherlands, this procedure led to an almost artificial landscape in which the human scale and community activities are almost nonexistent. The drainage of land, lakes, and parts of the sea has been highly successful, coursing land that has subsided under the average sea level, creating monotonous landscapes and resulting in an alarming loss of biodiversity. Before and during prolonged rainy periods, automatically driven pumps discharge water so efficiently that there is not enough water available during the summer, with severe consequences for crops, cattle, and nature. Water board management serves industrialized farmers who operate in a world market guided by economic benefits. To overcome these adverse effects of the scaling up and rationalization of the landscape, the *polder-boezem* system needs to be drastically rethought. This non-resilient approach to the landscape counts for many regions in so-called developed countries.

As can be learned from long-lasting water systems, the systems' fine-tuning and adaptability mode must be site-specific, flexible, and collaborative. The design should not be set up for productivity alone but should also integrate other aspects, such as the quality of the living environment for all kinds of living beings. However, with the perpetuation of today's consolidated individualistic lifestyle, it is not easy to influence humans' relationships with other beings in a more caring and meaningful way. However, experimenting with community involvement in relation to waterscapes has proven to have great results, especially when starting at smaller local scales. This is the case of the Lankheet Estate, where open-minded landscape management supported by active community involvement has positively impacted the estate over the past few years. This case proves that taking care of a water system means interacting with a living entity that requires constant support and attention, especially during rapid alterations. Thus, by adapting to climate change by using water in multifunctional ways and considering time holistically, water systems can be managed more sustainably.

Het Lankheet

Het Lankheet (Lankheet Estate), approximately 0.06 km^2 in size, is located between Haaksbergen and Eibergen in the Netherlands, near the border with Germany. The history of the estate goes back to 1188. The cultivation of woods, streams, moors, peat bogs, and fens, with its old farmyards, characterizes this landscape. The Buurserbeek, a tributary of the IJssel, flowed through the estate for a long time. In the Middle Ages, local farmers discovered that the water from the streams was rich in lime and minerals. To get these nutrients onto their land, a clever water system was developed consisting of ramparts, dams, and ditches guiding the nutrient-rich water to flood the land. This centuries-old agricultural technique called *graslandbevloeiing* improved the fertility of the soil while the water movement prevented the peat from freezing during the winter. In spring, the harvested grass was early and wealthy. Around 1400, the Buurserbeek was diverted to connect with the Schipbeek, making shipping feasible for the commercial exchange of wood, charcoal, and clay from Germany, butter and cheese from Holland, and eventually textiles from the region.

The Fall of Modernity

During the 20th century, the historical fertilization method was replaced by chemical fertilizers. Rather than overflow the meadows, water had to be drained from the grassland as quickly as possible so as not to flush away the chemicals. Scaling up the meadows became easy. Forests were cut and swamps were drained to reclaim even more agricultural land ready to be exploited with the help of fertilizers. The diversity of the estate and the water system subsequently disappeared. Groundwater levels then dropped significantly, resulting in a consistent drying out of the landscape on a large scale.

Periodic nuisance and rainwater increased while the accelerated drainage caused large volumes of water that the system could not handle. Previously, these volumes were stored in swamps acting as natural overflow areas and, when needed, used for irrigation. Moreover, in modern agriculture, little attention is paid to the role of soil organisms, such as bacteria, fungi, and worms, with the focus being placed on manure and herbicides for the crop. An example of lost knowledge is that a rich soil life provides nutrients for a prosperous, growing yield. Due to the increasing use of slurry, chemical fertilizers, and poisons, soil organisms and insects have decreased dramatically. The soil structure also deteriorated and compacted due to the use of heavy machinery on the land. As a result, less water is retained, more greenhouse gases are emitted, and less CO2 is stored. Due to the rationalization of farming, fewer and less knowledgeable people were needed to work in the fields.

A Socio-Ecological Estate

At the end of the 20th century, the relevance of landscape and ecological thinking toward wiser water management and nature restoration was finally recognized due to accelerating climate change. As supported by research on traditional water systems conducted as part of the Circular Water Stories research, this entailed turning the gaze to local community-based practices with a particular focus on water realms (irrigation, storage, etc.).[1]

In 1999, part of the historic flowing meadows of Lankheet Estate was restored, and today, 0.14 km^2 of meadow has been converted into natural flower-rich grassland thanks to traditional irrigation methods. Slurry and fertilizer are no longer used on these hayfields since the fields are once again flooded by the streams. According to different watering schedules for winter, spring, and early summer, freshwater flows over the land in a controlled manner, residing for four to six weeks. This stimulates soil life, efficiently converting the excess nitrogen and phosphate and producing flower-rich grassland. Managing and maintaining the system of ditches and locks using the slightest differences in height requires a lot of care and attention throughout the process. Traditional grassland irrigation allows farmers to produce high-quality clippings with minimum fertilizers while maintaining a maximum result in terms of natural values. Since periods of drought now last longer, this technique of slow water infiltration becomes even more valuable. This intangible heritage thus contributes to facing contemporary climate challenges.

To restore the adaptability of the water system, Strootman Landschapsarchitecten, together with the Research Center of Wageningen University (Plant Research International, Alterra), learned from the past. This collaboration started in 2004 and fostered adaptation to rapid climate change challenges—such as droughts, biodiversity loss, pollutant seepage, percolation, etc.—through century-old practices while prompting the conversion of Lankheet into an *ecological estate.* By restoring the traditional irrigation method, a water purification plant was introduced to filter the Buurserbeek flowing through the estate and ending up on the fields. To increase the water quality, 18 interconnected basins have been constructed. Within these, reeds perform water filtration through absorption, functioning as helophyte filters and removing excess phosphate and nitrogen. The purified water then flows into beautifully shaped mirroring ponds and then to the restored medieval streams, which run through the historic meadows. Clean water slowly infiltrates the ground, which helps the vitality of the forest, which is more susceptible to insect pests and fungi due to increasing summer droughts. The entangled system of watercourses, reed ponds, storage basins, water meadows, forests, and other landscape interventions (e.g., sculptures and bridges) enlarge the human and non-human relationships with the landscape and the water system.

To let the meadows overflow, the community needs to become active. Ditches are dug, partly by a light machine and partly by hand, for the water to reach the fields. A sluice at the beginning of the irrigation ditch can regulate the amount of water flowing into the fields. When the water runs into the ditch, people block the water flow with a board on a stick, guiding it to run over the fields. This process is repeated until the entire lot is wettened.

Current Role of the Lankheet Estate Water Community

The transformation of Lankheet into a more resilient and sustainable estate was only possible with the involvement of a newly established water community made up of people and organizations.[2]

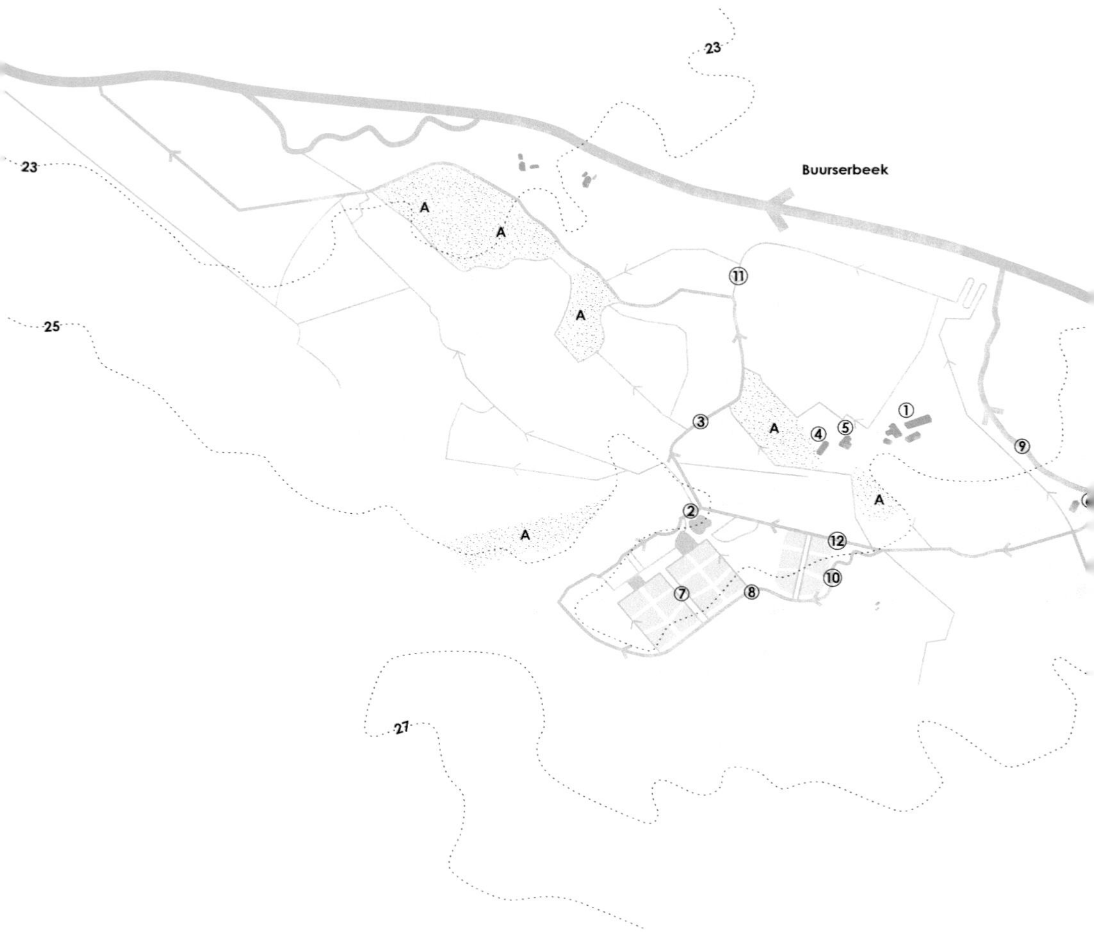

Figure 6.1 | Water flows of Lankheet Estate. The estate is configured as a continuous alternation of open meadows for hay cultivation and dense forest for timber production. Following the terrain topography, water is conveyed within Het Lankheet, flowing through the reedbed purification system and then again in the waterways. The water meadows are close to seasonally flooded canals. Map conceived by Inge Bobbink and Amina Chouairi and realized by Amina Chouairi.

Figure 6.2 | Lankheet Estate reed fields and water meadows. The zoom presents the complexity of the reeds filtering basins and how the purified water is directly employed to flood the traditional water meadows. In the 18 filtering ponds, reeds absorb pollutants (carbon, phosphorus, nitrogen, pathogens, and toxins) in the Buurserbeek water. Additionally, flourishing duckweeds contribute to pollutant absorption. The water meadows, flooded from twice to three times per year, produce the fodder harvested in late spring. Map conceived by Inge Bobbink and Amina Chouairi and realized by Amina Chouairi.

Legend

Figure 6.1

A | Watermeadows
1 | Small farm
2 | Landscape artwork
3 | Aqueduct
4 | Fieldwork centre
5 | Farm De Pingel
6 | Timber yard, drying shed and hut
7 | Purification filters
8 | Belvedere
9 | Lock
10 | Bridge
11 | Bridge-dam
12 | Waterpark entrance

Figure 6.2

A | Watermeadows
1 | Waterpark entrance
2 | Bridge
3 | Belvedere
4 | Purification filters
5 | Landscape artwork

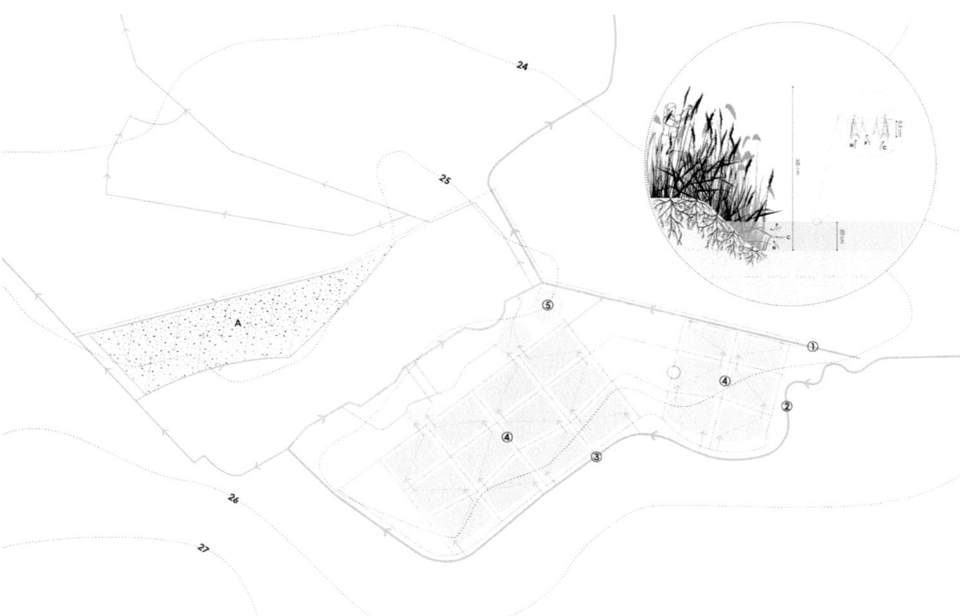

While still part of Lankheet Estate, the Lankheet Water Park is managed by a charitable foundation (*Algemeen Nut Beogende Instelling Foundation*): *Stichting Waterpark en Veldwerkcentrum Het Lankheet*, in short, the Waterpark Foundation. As stated by the Foundation itself, the main objective of the Waterpark Foundation is 'to promote and manage the Waterpark on the estate and the associated *Veldwerkcentrum*. [...] the Waterpark Foundation generates resources to realize its charities, such as care and culture projects and scientific research into multiple sustainable land uses, including nature development and education/public information. All this with the theme of water as a guideline' (Landgoed Het Lankheet Waterpark, 2022). All together—the Waterpark Foundation, the Het Lankheet Fieldwork Center (*Veldwerkcentrum het Lankheet*), the Association of Natural Monuments (*Vereniging Natuurmonumenten*), and the Kempen and Central Limburg Management Unit—have added the *Traditionele Bevloeiing van Grasland* (i.e., the traditional grassland irrigation) to the Inventory of Intangible Heritage of the Netherlands.[3] Several organizations are involved in the traditional irrigation of grassland: Lankheet Estate (*Natuurschoonwet BV*), the land owner of one of the sites, supporting landscape restoration and design; De Groene Waaier Foundation, a partnership of local managing organizations—the State Forestry Department (*Staatsbosbeheer, Haaksbergerveen*), the Natural Monuments (*Natuurmonumenten, Buurserzand and Witte Veen*) and Lankheet Estate (*het Lankheet, Assink and Aaftink*); the Water Board Rijn en IJssel, responsible for water management and water supply in the catchment area. These organizations are jointly accountable for restoring historic water systems, purification parks, and retention basins. The Historical Circle Haaksbergen, the Association for Nature and Environmental Education (IVN), and *De Schipperskring* provide volunteers for the seasonal irrigation practice. Simultaneously, the *De Pelterheggen* team manages, maintains, and protects the irrigation system.

This heterogeneous yet cohesive system of actors and stakeholders has agreed on a set of *Borgingsacties* (i.e., safeguarding actions) to manage and maintain the traditional irrigation system. Between 2021 and 2024, these actions will collide in providing teaching packages, contributing to community learning, developing material for primary education, organizing exchange trips, and involving adjacent farmers in practical activities. After this period, the status of Lankheet Estate will be reviewed by the Knowledge Center for Intangible Heritage Netherlands (*Kenniscentrum Immaterieel Erfgoed Nederland*). This set of activities is not only the result of a successful socio-cultural mechanism around the landscape but also directly contributes to the creation, consolidation, and expansion of a tight water community around Lankheet Estate.

Embrace Dynamics and Changes

For more than eight centuries, Lankheet's primary use was productivity. Timber has been produced since the 14th century, agricultural fields have provided wheat, barley, and maize, and rolling pastures have provided fresh grass for animals. Due to the adverse effects of farming rationalization and insufficient water availability, the estate commission began collaborating with Wageningen University, which has proven to provide farsightedness while tackling challenges and issues. Although Lankheet Estate is a historic landscape, this does not imply that time has stood still; on the contrary. As described, many organizations and people are involved in making that happen. Notably, a great deal of knowledge relating to the old systems had to be reinvented. New parts, such as the filters, had to be added to meet modern environmental conditions. Nevertheless, the traditional water system has developed a new meaning—one that is not limited to those who live and work in the estate. By widening the spectrum of activities within the public program of the estate (e.g., walking, cycling, horse riding, etc.), more and more people were reached, and became involved. Lankheet's restored water landscape, along with its new water community, evolves dynamically in terms of its geographical and historical contexts and scales. This continuous adaptability of flows is 'the key to the existence and sustainability of water communities' (Tanabe, 1994, cited in Thaitakoo and Shaw, 2010, p. 265).

Lankheet addresses the growing 'demand for more vital approaches offering new possibilities for narrative, experience, collaboration, and design' (Kolen, Renes and Hermans, 2015, p. 23). As presented in *Landscape Biographies*, heritage—in the form of long-lasting water systems—should not be seen as a limiting obstacle but rather the primary source of inspiration for creative responses to moments of crisis. The estate's landscape and heritage management approach has proven to its long-lasting functioning and life rather than its musealization. The Lankheet Estate project has proven that virtuous water management through the involvement of people is vital in creating a sustainable, ecological, and culturally rich landscape. Labor-intensive maintenance can become a meaningful use of time and restore human relationships with nature. Based on the Lankheet model, changes in the Dutch polder landscape and many other water systems worldwide can also be imagined.

Figure 6.3 | Lankheet Estate Water Community. Digging the ditches that bring the water to the field. Photograph by Ari Vos.

Figure 6.4 | Lankheet Estate Water Community. Stopping and guiding the water flow by using a board. Photograph by Jan Dolfing.

Notes

1 | The Circular Water Stories investigation refers directly to the concept of 'landscape biography,' a research approach that describes and maps human influences in (urban) landscapes. The idea was reconsidered and further developed by Bobbink and Loen in the publication *Visual Water Biography* (Bobbink and Loen, 2020).

2 | 'Community' from the Latin *commùnitas* (society, participation) a derivative of *commùnis* (performing one's duties together), a derivative of *munus* (obligation, but also gift), with the prefix *cum-*. *Munus* is the obligation, the function: *Communis*, in its smallest terms, describes the character of the person or thing that performs tasks in a shared situation together with others. It is from this substratum that *commùnitas* emerges as a relationship of commonalities. It appears in the goal of a shared burden, splendidly ambiguous because *munus* is the obligation but also (and according to some, it is the most ancient meaning) the gift, the favor, the offering in the vote.

3 | The 'traditional grassland irrigation' practiced in Lankheet Estate has been added to the *Inventaris Immaterieel Erfgoed Nederland* in April 2021. The Inventory of Intangible Heritage of the Netherlands contains intangible heritage for which the heritage community (or a group or individual) has written an assurance plan to give the heritage a future. This plan has been tested by an independent review committee. Heritage care is evaluated every three years.

Acknowledgments

The photos of the water workers were made available by Eric Brinckmann, Stichting Waterpark en Veldwerkcentrum Het Lankheet.

References

Bobbink, I. (2016) *The Landscape Architecture of the Polder-boezem System: Structure and Form of Water Network, Water Pattern and Water Work in the Dutch Lowlands.* Delft, BK Books.

Bobbink, I., Chouairi, A. and Di Nicola, C. (2022) 'Visualizing water. Using the illustrative method to learn from long-lasting water systems.' *Blue Papers: Water & Heritage for Sustainable Development*, 1 (1), pp. 106–117. Available at: <https://bluepapers.nl/index.php/bp/article/view/5> [Accessed 24 August 2023].

Bobbink, I. and Loen, S. (2020) 'Visual water Biography.' *SPOOL Circular Water Stories*, 7 (2), pp. 5–21. Available at: <https://spool.ac/index.php/spool/article/view/153> [Accessed 7 December 2022].

Circular Water Stories (2020) *Circular Water Stories: Landscape Architecture Research and Education.* Available at: <https://circularwaterstories.org/> [Accessed 24 October 2022].

Kolen, J., Renes, J. and Hermans, R. eds. (2015) *Landscape Biographies: Geographical, Historical and Archaeological Perspectives on the Production and Transmission of Landscapes.* Amsterdam, Amsterdam University Press.

Landgoed Het Lankheet Waterpark (2022) *Over het Waterpark.* Available at: <https://hetlankheet.nl/waterpark/> [Accessed 28 November 2022].

Thaitakoo, D. and Shaw, R. (2010) 'Essentials of water communities and its future perspectives.' In: Waugh, W.L.Jr. ed. *Water Communities: Community, Environment and Disaster Risk Management.* Bingley, Emerald Group Publishing Limited, pp. 263–274.

Tri Prestasia, A. and Kim, B. (2020) 'Living with nature: Water stories of Kampung Naga.' *SPOOL Circular Water Stories*, 7 (2), pp. 59–75. Available at: <https://spool.ac/index.php/spool/article/view/162> [Accessed 20 December 2022].

7 | Resilient Design for an Altered Planet

Living Breakwaters for a More-Than-Human (Co)Design

Philippa Brashear

Design Project Data

Location
Tottenville | Staten Island | New York | USA | North America
Project Team
Lead agency | New York State Governor's Office of Storm Recovery
Design and engineering team | SCAPE Landscape Architecture (Prime) | COWI | ARCADIS | SeArc Ecological Marine Consulting | WSP | MFS Consulting Engineers | Prudent Engineering | LoTEK
Environmental and permitting | AKRF
Physical modeling | National Research Council of Canada
Construction Managers | Ramboll, Baird
Construction Contractor | Weeks Marine
Date
Design | 2013 | 2019
Construction | 2021 | 2024
Coordinates
40° 30' 45.966" N | 74° 15' 7.093" O
Area
n.a.
Length
730 linear meters of breakwaters along approximately 1.6 km of shoreline
Investment
Construction | US $78.7 million
Total | US $107 million

Living Breakwaters

Eight years ago, SCAPE published *Towards an Urban Ecology* (Orff, 2016), a 'manufestograph' for their practice. This publication introduced tools to help designers and citizen activists conceptualize their local environment in new ways and inspire them to sustainably recreate the built-natural world. These tools include a design framework for reviving ecosystems and landscape identity, which enable us to cohabit on this planet by extending our design thinking beyond our own species and engaging people in productive dialogues that engender creative collaboration. Living Breakwaters—already an award-winning concept—was the capstone of the book, an illustration of how these concepts and ideas can be combined and scaled to effect positive change for communities and ecosystems. At the time, the project was still in the conceptual phase. Fast forward eight years to today and the project is finally under construction. The process of taking this vision to reality has been long and often challenging—practice is hard work—but it has also taught us new lessons about what it takes (and will take) to create the real change required for successful adaptation. This essay explores how the Living Breakwaters have helped refine and articulate principles for ecological design in practice: How designers can revive ecosystems and landscape identity, enable people to co-exist with their natural environment and inspire action, as well as help people physically adapt to a changing world. It has also shown that these efforts can be scaled to create positive change, and indeed hope, in a resilient, climate-adapted future.

Rebuild by Design as a Catalyst for Change

In October of 2012, Hurricane Sandy pummeled the New York metropolitan region, impacted 13 states, and cost over 186 lives as well as more than US $65 billion in damages and economic loss. We were not prepared, and it was a wake-up call to the climate change-induced threats posed by sea level rise and higher-intensity storms in our region. The recovery and rebuilding effort in the wake of the storm was massive, undertaken at the local, state, and federal levels and catalyzed by a large influx of federal disaster recovery funding. Most of this funding was allocated according to traditional criteria and through existing programs that place a heavy emphasis on rebuilding.

Notably, it is far easier and faster to obtain funding to rebuild a home or a piece of infrastructure where it was and the way it was than it is to fund adaptation or relocation. And yet, in the United States, this is the way that most flood risk reduction and climate adaptation projects occur—funded and designed in the wake of a catastrophic event. It usually takes a disaster to create the opportunity to adapt. Few cities or states in the impacted region had climate adaptation plans or coastal flood risk management programs in place when Hurricane Sandy hit, and there was no comprehensive federal policy or plan for proactive climate adaptation of flood risk management. Since the land-use decision-making necessary for effective adaptation and water management rests almost entirely at the local level, it is challenging to plan—much less implement—the coordinated measures necessary for effective flood management and climate change adaptation, with a notable exception being the United States Army Corps of Engineers (USACE), which designs and builds water resource projects targeted at improving navigation, reducing flood risk, and restoring aquatic ecosystems; however, these activities are frequently siloed from each other. USACE projects must be authorized by the US Congress, require local cost sharing, and include an extensive feasibility study involving a very specific benefit-cost analysis and evaluation criteria. These factors often make it challenging to effectively respond to local and regional priorities, innovate, or advance multi-functional projects. It also means they take a long time to complete.

In the wake of Hurricane Sandy, most funding went to traditional recovery and rebuilding efforts. However, President Obama's Hurricane Sandy Rebuilding Task Force, led by Housing and Urban Development (HUD) Secretary Shaun Donovan, also launched an innovative design competition and set aside US $1 billion of HUD Disaster Recovery Funding to go toward proposals that came out of the competition. Rebuild by Design invited applications from design-led, interdisciplinary teams. It ultimately selected eight teams to engage in a multi-stage competition to develop implementable strategies for climate resilience in the Sandy-impacted region. Rebuild by Design was a facilitated process that foregrounded design thinking and involved in-depth research, deep cross-disciplinary collaboration, ongoing dialogue with stakeholders, and iterative design. While funding for the projects came from HUD, funding for the competition came from the Rockefeller Foundation and other philanthropic sources. The competition brought together government, technical experts, creative thinkers, and stakeholders in a collaborative setting, investing in the beginning of the design process and supporting a transparent process of research and engagement. In the words of then-Director of Rebuild by Design Henk Ovink, this 'was a sidestep away from the institutional world. A detour around negotiations, the process aimed to build understanding and trust' (Bisker et al, 2016, p. 24); 'Rebuild by Design is not about making a plan, but about changing a culture' (Bisker et al, 2016, p. 10). It was a fundamentally different way for the federal government to respond to a disaster and was a catalyst for the collaboration, creative thinking, and innovation that the competition generated.

Living Breakwaters

Our SCAPE team approached the competition with the premise that in a region as inextricably tied to the water as New York, we require adaptation strategies that allow us to maintain our connections with the water—physically, ecologically, socially, and economically—rather than walling the water off. Our premise was that we could develop place-specific adaptations that combine risk reduction with ecosystem improvement and social resilience, applying a layered approach to reduce risk and foster a culture of resilience.

The outcome of our participation in Rebuild by Design was Living Breakwaters, a proposal for a necklace of nearshore breakwaters—partially submerged structures built of stone and ecologically-enhanced concrete units—to break waves, reduce (and eventually reverse) erosion, provide a range of habitat spaces for oysters, fin fish, and other marine species, and foster social resilience through education and stewardship. HUD awarded New York State US $60 million to design and implement Living Breakwaters. Thus, in 2015, we embarked on the process of making this vision a reality with the state.

That same year, SCAPE published *Towards an Urban Ecology* (Orff, 2016). In this 'manufestograph' for our practice, we laid out strategies for designers and citizen activists to conceptualize their local environment in new ways and inspire them to sustainably remake the built-natural world. Four key principles serve as the framework for this book and our stance and practice: *Revive*, *Cohabit*, *Engage*, and *Scale*. Living Breakwaters was the capstone project featured in the book. However, it only remained a concept at the time—a bold project for climate adaptation with the hope, but not certainty, of being physically realized.

Figure 7.1 | Competition phase section-perspective rendering of the Living Breakwaters system. Image by SCAPE Landscape Architecture.

CONSTRUCTED TIDE POOL
HARD CLAM
TREET
ater rock
uvenile fish,
els.
UPLAND ISLAND
Exposed island habitat free from predators can be used by seals and birds.
SUBTIDAL ROCKY SUBSTRATE
Subtidal shallow water rock enhancements for juvenile finfish, lobsters, and shellfish.
MUDFLATS
Zones of moderate sedimentation create habitat for hard clams, benthic fish, and eelgrass.

Today, nine years later, the project is finally under construction. The process of taking this vision to reality has been long and often challenging—practice is hard work—but it has also served as a proof of concept. The process of designing and constructing the Living Breakwaters has enabled us to articulate our principles for ecological design through the active work of designing and building landscape infrastructure. It has also given us confidence to believe that we can create positive change, and indeed hope, in a resilient, climate-adapted future.

Revive Landscape Systems

The principle of *Revive* involves realizing the place-making potential of natural systems. In our contemporary environment, this calls on us to renew, relink, and regenerate the structure and function of natural systems when we can no longer restore them. In the context of Living Breakwaters and similar nature-based coastal adaptation projects, this involves determining how to work with—rather than attempt to control—geologic, hydrodynamic, and aeolian processes.

Tottenville, on the south shore of Staten Island where the Living Breakwaters are being constructed, was once surrounded by extensive beds of *Crassostrea virginica*, the eastern oyster. Oysters were so plentiful that Tottenville has been referred to as 'the town the oyster built' because their harvest was so critical to the local economy in the 1800s. These reefs provided habitat for migratory fish and other species as well as food for the Lenape people indigenous to the area, and then later for European and African settlers (MacKenzie Jr., 1996). They also sustained the beaches and marshes along the shore by buffering them from erosive wave action. These reefs no longer exist—destroyed by over-harvesting, poor water quality, and sedimentation associated with dredging, oysters and their reefs are scarce in the harbor today. Their loss has meant the loss of a natural coastal defense system for dampening wave energy and keeping erosion in check (Temmerman et al, 2013; Brandon, 2016). Now highly exposed to sediment-moving wave action, the Tottenville shoreline has been eroding over time, leaving the homes, infrastructure, and ecosystems on shore more exposed to erosion and the damaging waves that come with coastal storms.

We knew early on that we could not re-create the reefs that once existed; that humans had so heavily altered New York Harbor as to render the reefs that thrived in the bay of the 1700s unsustainable. However, we posited that we could design something to revive the ecosystem functions those reefs provided, including breaking waves and managing shoreline erosion. Not something that would keep the water out or stop the beaches from moving, but something that would reduce the most damaging storm impacts (wave attack) and chronic beach erosion. The design challenge we faced was how to achieve this not by directly manipulating the coast, but rather by manipulating the processes that shape it. In concept, breakwaters were the appropriate form of infrastructure to do this since they are traditionally used to reduce wave action and reduce on-shore erosion. However, to move from concept to constructible design, we had to develop the configuration, length, spacing, height, and cross-section for the breakwaters that would generate the on-shore conditions we sought to achieve.

This was achieved through collaboration with coastal engineers and modelers to understand the site through the forces that shape it, develop prototypes for breakwater designs and configurations, and then iteratively test and refine those designs under various future scenarios (defined by factors such as sea level rise, water levels, and wave angles). Everyone involved stepped outside of their disciplinary silos to contribute the design concepts, criteria, and assumptions that would guide the testing and evaluation. Designs were iteratively prototyped and modeled to understand their potential interactions with waves, currents, and sediment: Shoreline change models to simulate the gradual movement of the shoreline, wave models to test storm events, sediment transport modeling to ensure that we were not creating unanticipated negative effects, and physical modeling to validate the structural stability of our design. This process required designers, ecologists, engineers, modelers, and stakeholders to develop a shared language and means of communication around design performance and model results to confirm a shared understanding of what we were seeing—outcomes and uncertainties—and enable honest and productive collaboration.

The idea that any coastal infrastructure can eliminate risk is a dangerous fallacy. The geologic and hydrologic processes that shape our shores cannot be overcome, and our efforts at coastal adaptation will surely fail if we rely solely on infrastructure designed to keep water out, fix and harden shorelines, and stop sediment movement. Like reefs, the Living Breakwaters and other forms of nature-based infrastructure do not keep water out; instead, they help us manage risk by working with—rather than fighting against—water. The beaches and dunes will shift with the seasons, but they will not be lost; buildings on shore will need to adapt or flood with severe storms, but they will not be knocked off their foundations.

To design this type of nature-based infrastructure as landscape architects, we need to expand our professional collaborations and engage in a rigorous, science-based, and scenario-driven process of testing and refining our designs. It will also require us to acknowledge the unknowns inherent in modeling projected but uncertain future conditions and adapt our representation techniques to effectively enable honest dialogue and decision-making around managing risk.

Figure 7.2 | Setting up physical modeling of the Living Breakwaters at the National Research Council of Canada facilities in Ottawa as part of the design process. Image by SCAPE Landscape Architecture.

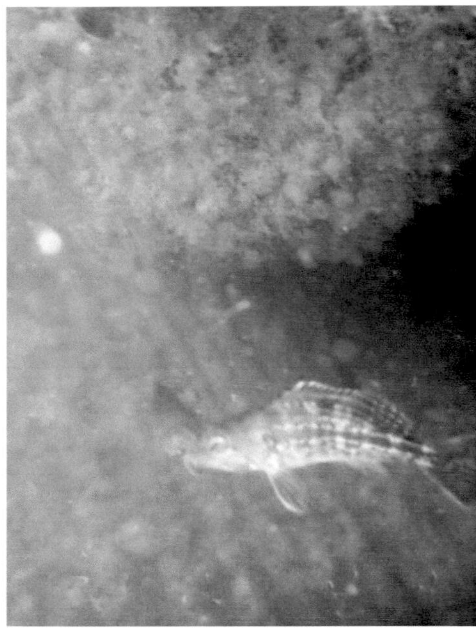

Figure 7.3 | Wildlife already being observed on the Living Breakwaters: comb jelly (*Mnemiopsis ieidyi*). Photograph by COWI Engineers.

Figure 7.4 | Wildlife already being observed on the Living Breakwaters: juvenile Black sea bass (*Centropristis striata*). Photograph by COWI Engineers.

Cohabit Our Changing Planet

The principle of *Cohabitation* calls on us to extend our design thinking beyond our own species and create environments 'supportive of life and growth [for all] ... in which a diversity of species can live together and prosper on mutually beneficial terms' (Orff, 2016). As designers, we must define other species as stakeholders in our designs from the outset, integrate their needs into decision-making throughout the design process, and deliver infrastructure that improves—rather than detracts from—their well-being.

The oyster reefs that once existed offshore of Tottenville were the cornerstone of a vibrant aquatic ecosystem: As food and water filters (oysters) and habitat (oyster reefs), they supported life and growth for a diverse array of aquatic, avian, and terrestrial species as well as the livelihoods of many people (Grabowski and Pterson, 2007; MacKenzie Jr., 1996). They have been a philosophical and physical inspiration for Living Breakwaters since we sought to design a thriving ecosystem with infrastructure that would not only support human well-being but also enhance biodiversity and enable many species to prosper. This began with understanding the inhabitants of Raritan Bay and their needs: Juvenile migratory fish who require habitat that provides shelter during this vulnerable life stage, oysters and other sessile species that could organically add to the structures, other invertebrates such as crabs and lobsters, as well as birds and marine mammals who feed on aquatic species.

The Living Breakwaters are designed to provide habitat for these users and support a healthy, thriving local ecosystem. Made from a combination of natural stone and ecologically-enhanced concrete units, the Living Breakwaters are porous structures full of crevices and void spaces that allow water flow within the structure and provide aquatic species with foraging opportunities and refuge from waves, desiccation, and potential predation (Silva et al, 2008; Sherrard et al, 2016). Features we call 'reef ridges' are designed to mimic the spur and groove formations found in many naturally occurring reefs (Rogers et al, 2013; Roberts, Wilson and Lugo-Fernández, 1992). These features dramatically increase the intertidal and subtidal surface areas of the breakwaters and were modeled to confirm they would promote the appropriate water movement to transport larvae and flush sediments while minimizing scour. Artificial tidepools are included within the intertidal zone of the main trunk, the shallow-sloping tops of the reef ridges, and in small cut-outs in the crests of shorter breakwaters. By retaining water between tides, tidepools provide a unique habitat including important nursery grounds, feeding habitat, and refuge for a wide range of organisms (Firth et al, 2014; Perkol-Finkel and Sella, 2015).

Understanding the breakwaters' diverse potential users and developing and refining a habitat design specifically for these target users were informed by biological surveys of the area, local studies, peer-reviewed literature, and experts including marine biologists and ecologists as well as citizen scientists, fishermen, and other observers and stewards of the harbor. A deep understanding of the local environment, species needs, and biological interactions is required for effective ecological design; thus, so is collaboration with people who have taken the time to observe, research, and study the needs and behaviors of other species with whom we share this earth.

Supporting healthy ecosystems, biodiversity, and the well-being of other species is often treated as an aspiration rather than a commitment in climate adaptation projects: Typically referred to as 'co-benefits,' they are frequently presented as fortunate extras to risk reduction—a bonus we are lucky to receive rather than a fundamental element of what the project must provide. This must change. Notably, climate change is not the only human-induced crisis we face. We also face an extinction crisis that is the direct result of habitat destruction and fragmentation, overharvesting, and pollution. Designing to improve biodiversity and support healthy ecosystems is not optional; it must be a fundamental part of the process and product of our designs. Since healthy ecosystems are critical infrastructure, we must work to preserve those that remain. However, we must also build forms of infrastructure that function as ecosystems.

Engage People to Opt in to Adaptation and Stewardship

Engaging stakeholders in the planning and design process is imperative for social responsibility, environmental justice, and the very success of projects. The principle of *Engage* is about generating new community formats around a shared purpose and integrating education. To be effective, climate adaptation projects must go beyond traditional engagement in the design process: the project itself must engage people to opt in to undertaking behavioral changes that will improve their own and society's ability to thrive in the face of climate change. Education and social change are part of this adaptation.

For Living Breakwaters, we engaged a wide range of stakeholders throughout the design process. More importantly, we have generated the social infrastructure required to raise people's awareness of natural systems and how we rely on them, increase their capacity for adaptation, and involve people in the stewardship of their environment. The core of this social infrastructure has been youth education and community stewardship advanced through a partnership with the Billion Oyster Project (BOP), a NYC nonprofit 'restoring oyster reefs to New York Harbor through public education initiatives' (Billion Oyster Project, 2022). BOP was founded in 2014—the same year that we were beginning to develop Living Breakwaters—with the philosophy that 'restoration without education is temporary,' and the keen observation that long-term efforts such as the successful restoration of our harbor ecosystems rest in the hands of those who are now youth. BOP offers public school students, volunteers, and community scientists the opportunity to learn about New York City's rich oyster history and participate in the movement to restore it. While we had a vision of the bay/harbor as a classroom, BOP was making it a reality. Since 2014, they have restored 100 million live oysters to New York Harbor and engaged over 8,000 NYC school students from 100 schools. Just as we needed to partner with engineers and biologists to design physical infrastructure and habitat, we also needed to partner with educators and active community stewards to develop the social infrastructure.

From the outset, the question asked was: 'How can the Living Breakwaters project and the team we have assembled help you realize the educational and stewardship outcomes we both agree are needed?' During the competition phase in 2014, the team helped create the Staten Island Oyster Gardening Manual and held public workshops where teachers signed up to become BOP schools. The competition delivered not only our design proposal to HUD but also a concrete, usable education tool developed in collaboration with educators based on their needs. Even in that early phase, the project generated something tangible and useful for harbor stewards.

When the project was awarded to New York State, the Governor's Office of Storm Recovery (GOSR) took on not only the physical project and contract with the technical team, but also committed to building the social infrastructure, including an agreement with BOP to expand their educational program in Staten Island, develop concrete usable educational materials, scale up their shell recycling program, and eventually be responsible for restoring and monitoring oysters on the breakwaters. The state also invested in social infrastructure.

In 2015, BOP had no participating schools on Staten Island. Since 2019, they have engaged 33 schools and over 2,000 students on Staten Island in their programming. Supplementing BOP's main programs, the project team and BOP hosted shore walks, educational installations, monitoring days, and even pilot oyster installations to test and monitor techniques that would be used on the breakwaters. These events provided hands-on engagement with restoration, stewardship, and different aspects of the project. SCAPE also helped BOP develop an online curriculum using the Living Breakwaters as a case study. Launched in early 2020, the curriculum was available during the COVID-19 pandemic, when access to online learning and outdoor education was suddenly extremely important. This will not end with construction: When the breakwaters are complete, BOP is funded to install and monitor oysters on the breakwaters and continue their educational programming.

Awareness, education, and stewardship are foundational to climate adaptation. Without building the social infrastructure required for change, our physical infrastructure—nature-based or not—will fail. We need to help cultivate and support advocates for action who are aware of the urgency, stakeholders who understand that adaptation will require change and a generation of leaders and stewards who commit time and money to maintenance and management. As designers traditionally responsible for physical projects, we can and must also apply our skills in design communication to awareness raising and education. We must invest in engagement and education as a part of all climate adaptation work and ask our collaborators and clients to do the same.

Scale Positive Outcomes

Scale and *scaling* are about contributing to and catalyzing positive change beyond the bounds of a particular project. It is the synthesis and combined application of the other approaches of revive, scale, and cohabit with attention to the larger context in which we need to effect change. The climate and biodiversity crises are immense, and they will require collective societal action. As landscape architects, being able to design and implement physical projects at a defined scale is critical to effect change through our practice. However, our work must engage larger environmental and social systems, while the benefits must reach far beyond the individual project or site.

The Living Breakwaters exist at a particular scale in the landscape—a little over 1.6 Km of coarse sandy shoreline, approximately 28,000 m^2 of nearshore waters in Raritan Bay (lower New York Harbor), a neighborhood in south Staten Island, New York City. However, the reach of their ecological and cultural benefits is far greater. Through the revival of hydrodynamic interactions, the creation of habitats that spawn biological activity, and the expansion of social and educational networks, the breakwaters will effect positive change far beyond the scale of their physical construction. The Living Breakwaters have also generated countless community conversations, academic papers, presentations, television segments, news articles, and other conversations. The project has created a platform for conversations that raise awareness of climate change and habitat loss, articulate the need for resilient ecological design, and call attention to reforms in the policies, regulations, and funding needed to remove barriers to similar projects being implemented. We hope that these will have ripple effects not only for the practice of landscape architecture but also for the design and engineering of infrastructure, policy, and society more broadly.

Scaling also requires us to commit ourselves and our work to a process of learning and adaptation. As humans, we are adept at re-making land and water for our own aims; however, to date, we have too often done this at the expense of other organisms and the environment. When we design and construct landscapes, we are intervening in a dynamic environment where the outcomes of intervention are never certain—and with climate change they are becoming even less so. This is particularly true in our coastal environments. In designing Living Breakwaters, we studied historic processes, observed the present conditions, and used the best available science to model our designs projected into the future. However, the work will not end with construction. Integral to the project is a monitoring program and adaptive management plan. The physical change in the shoreline as well as the biological activity on the breakwaters will be studied over time, thereby enabling us to understand and learn from how they function over time and make adjustments if needed.

In our rapidly changing environment, monitoring, adapting, and learning from our work is critical. We cannot afford to generate more negative impacts. Moreover, in our altered environment, we can only learn so much from the past. Thus, we need to embrace uncertainty and eschew the idea of our designs as fixed products. We must consciously and conscientiously consider and respond to the consequences of our work over time, modeling potential futures in the design process while also monitoring outcomes and applying the lessons learned.

Hope in Resilient Ecological (Co)Design

Where oyster reefs once existed in Raritan Bay, the Living Breakwaters are now emerging from the water. A year into construction, the completed portions are already breaking waves and attracting an array of species: Rocks are encrusted with barnacles, invertebrates inhabit tidepools, and sponges and other sessile organisms cover armor units. Above water, the breakwaters teem with roosting birds, and even bald eagles and harbor seals have been spotted. They represent a form of resilient, ecological infrastructure—the manifestation of these principles put into practice. A practice of (co)design.

The Living Breakwaters have taken a long time to transform from concept into physical form—nearly a decade. Over that time, it has been sustained by the dedication of an interdisciplinary project team, a committed owner, and the continued advocacy and support of stakeholders from many different walks of life. It has also required intense collaboration. The climate and biodiversity crises are larger than we can comprehend—much less address—on our own. Therefore, as designers, we must engage deeply with others who study the processes that shape our physical environment, who understand the species that inhabit it and their complex interactions, who manage or regulate the landscapes and waters in which we live, and who live, work, and teach in the communities where we work. These diverse voices and perspectives need to have an active place in our design process, and we must develop and hone our communication tools and methods of working to enable these critical collaborations to occur.

Figure 7.5 | Pilot oyster restoration being installed near the site by BOP staff, project team members, and volunteers. Photograph by SCAPE Landscape Architecture.

Figure 7.6 | The Living Breakwaters under construction. Photograph by SCAPE Landscape Architecture.

Figure 7.7 | The Living Breakwaters. Photograph by SCAPE Landscape Architecture.

However, (co)design must go beyond collaboration. It requires commitment: commitment from us as practitioners, from our clients, and from our stakeholders to understand and work with the natural and cultural processes that shape our landscape (*Revive*). It also requires a commitment to acknowledge and act on our shared kinship with all living things (*Cohabit*), to create social infrastructure (*Engage*), and to generate ecosystems for positive environmental change (*Scale*). We must be creative, but also conscientious. In practice, this means following through on the principles described here, laying out the outcomes needed to deliver on those commitments, developing and employing the means and methods to rigorously test our designs against them, jettisoning or revising concepts that do not stand up, and raising up and refining those that do. It also means being part of a larger movement. We must acknowledge that our work does not end with design or even implementation and that by working with natural processes and within ecosystems we are committing to an ongoing process of engagement, learning, and adaptative management. The Living Breakwaters project has been bold enough to inspire but pragmatic enough to become.

It can be difficult to maintain hope amid the global climate crisis. It can be even harder to design in the face of massive uncertainty and dire predictions. But we must. For me, as a principal at SCAPE, hope comes from actively applying the principles of revive, cohabit, engage, and scale in our work, and continuing to evolve our practice toward resilient ecological design. This is not a wistful, wishing, tentative hope, but rather the type of active, activist hope that writer, historian, and activist Rebecca Solnit described when she wrote:

> 'The hope I am interested in is about broad perspectives with specific possibilities, ones that invite or demand that we act. … You could call it an account of complexities and uncertainties, with openings. … Hope locates itself in the premises that we don't know what will happen and that in the spaciousness of uncertainty is room to act.'
>
> (Solnit, 2016, pp. xv–xiv)

I began this essay by describing the origin and catalyst for the Living Breakwaters project in the wake of Hurricane Sandy, highlighting one of the critical challenges to achieving climate adaptation in the United States—the lack of holistic climate adaptation plans and sustained funding sources to implement them. Over the past decade, we have seen a shift. Now, most states and major cities have climate adaptation plans as well as mitigation plans and policies. Moreover, the USACE is promoting Engineering With Nature (R) and recently published guidelines for nature-based infrastructure. Additionally, movements for a Green New Deal and a Blue New Deal have led to bold climate policies under the Biden administration. We have also seen unprecedented investment in climate adaptation since the Bipartisan Infrastructure Law has committed over US $50 billion to action to reduce risk from climate hazards, while the Inflation Reduction Act committed over US $370 billion toward clean energy and green infrastructure projects. In 2012, it took a disaster to catalyze and fund action. I have hope for more projects such as Living Breakwaters today because the context for proactive—rather than reactive—adaptation through resilient ecological design has finally arrived.

Acknowledgments

Thank you to the entire Living Breakwaters design team for your contributions to this work, the New York State for their commitment to implementation, and to the US HUD and New York State for funding. Furthermore, thank you to the team at SCAPE and Kate Orff for the articulation and application of these design principles. Finally, thank you to the staff at the BOP for your work and for providing information on your accomplishments for this article.

References

Billion Oyster Project (2022) *Our Story.* Available at: <https://www.billionoysterproject.org/our-story> [Accessed 10 September 2022].

Bisker, J., Chester, A., Eisenberg, T., Davis, S. and Ovink, H. eds. (2016) *Rebuild by Design.* Madison, WI, American Printing Co. Available at: <https://www.rebuildbydesign.org/digital-media/books/> [Accessed 1 October 2022].

Brandon, C.M., Woodruff, J.D., Orton, P.M. and Donnelly, J.P. (2016) 'Evidence for elevated coastal vulnerability following large-scale historical oyster bed harvesting.' *Earth Surface Processes and Landforms*, 41, pp. 1136–1143.

Firth, L.B., Schofield, M., White, F.J., Skov, M.W. and Hawkins, S.J. (2014) 'Biodiversity in intertidal rock pools: Informing engineering criteria for artificial habitat enhancement in the built environment.' *Marine Environmental Research*, 102, pp. 122–130.

Grabowski, J.H. and Peterson, C.H., (2007) 'Restoring oyster reefs to recover ecosystem services.' In: Cuddington, K., Byers, J., Wilson, W. and Hastings, A. eds. *Theoretical Ecology Series. 4 Ecosystem Engineers: Plants to Protists*, San Diego, Academic Press, pp. 281–298.

MacKenzie Jr., C.L. (1996) 'History of oystering in the United States and Canada, featuring the eight greatest oyster estuaries.' *Marine Fisheries Review*, 58, pp. 1–79.

Orff, K. (2016) *Toward an Urban Ecology.* New York, Monacelli Press.

Perkol-Finkel, S. and Sella, I. (2015) 'Harnessing urban coastal infrastructure for ecological enhancement.' In: Burgess, K. ed. *Proceedings of the Institution of Civil Engineers-Maritime Engineering Vol 168, Issue 3.* Amsterdam, Thomas Telford Ltd, pp. 102–110.

Roberts, H.H., Wilson, P.A. and Lugo-Fernández, A. (1992) 'Biologic and geologic responses to physical processes: examples from modern reef systems of the Caribbean-Atlantic region.' *Continental Shelf Research*, 12, pp. 809–834.

Rogers, J.S., Monismith, S.G., Feddersen, F. and Storlazzi, C.D. (2013) 'Hydrodynamics of spur and groove formations on a coral reef.' *Journal of Geophysical Research: Oceans*, 118, pp. 3059–3073.

Sherrard, T.R., Hawkins, S.J., Barfield, P., Kitou, M., Bray, S. and Osborne, P.E. (2016) 'Hidden biodiversity in cryptic habitats provided by porous coastal defence structures.' *Coastal Engineering*, 118, pp. 12–20.

Silva, A., Hawkins, S., Boaventura, D. and Thompson, R. (2008) 'Predation by small mobile aquatic predators regulates populations of the intertidal limpet Patella vulgata (L.).' *Journal of Experimental Marine Biology and Ecology*, 367, pp. 259–265.

Solnit, R. (2016) *Hope in the Dark.* Chicago, IL, Haymarket Books.

Temmerman, S., Meire, P. Bouma, T.J., Herman, P.M., Ysebaert, T. and De Vriend, H.J. (2013) 'Ecosystem-based coastal defence in the face of global change.' *Nature*, 504, pp. 79–83.

8 | Underwater Living Landscapes

A Landscape Plan for Calanques National Park

Miguel Georgieff

Design Project Data

Location
Calanques National Park | France | Europe
Project Team
Atelier Coloco | Andromède Oceanologie | Pierre David | Gilles Clément
Date
2015 | 2019
Coordinates
43° 12' 32.828" N | 5° 27' 9.485" E
Area
Marine area | 435 km^2
Terrestrial area | 85 km^2
Length
15 km
Investment
85,000 euro

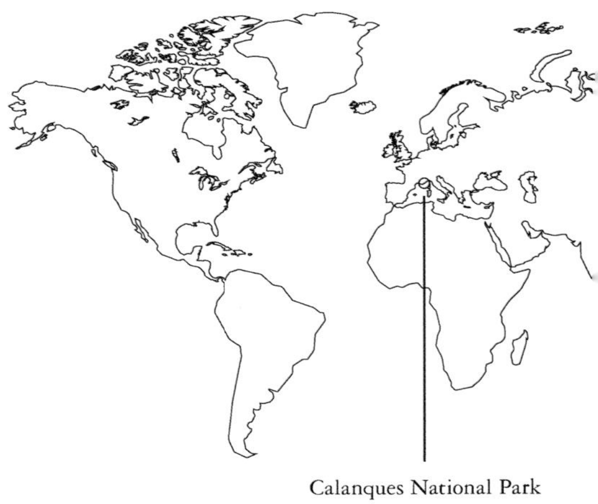

The Calanques National Park Underwater Landscape Plan (Parc National des Calanques et al, 2016) considers marine and terrestrial spaces as a single interdependent territory comprising exceptional landscapes. It constitutes the territory project's framework and nourishes the approach of a common vision between sea actors.

This multidisciplinary study proposes a vision of these landscapes, integrating marine, submarine, and terrestrial spaces to understand the issues and co-evolutionary relationships of different living environments.

To unite preservation efforts, this Landscape Plan bridges knowledge between experts (researchers, academics, biologists, etc.) and those who work the sea (fishermen, divers, artists, etc.). It provides a reflection on and representations of the roles of Mediterranean underwater landscapes in a global context. It defines common challenges of preserving these ecosystems following the impacts of direct or indirect actions on the environment.

This initiative is based on exchanges that question the notion of underwater landscapes to recognize a little-practiced common territory often reserved for experts, with a strong fragmentation of knowledge between specialties and categories of actors. This logic must consider underwater landscape evolution over immediate and long-term time frames to evaluate actions and their effects over time while respecting natural dynamics as the basis for a territorial project combining marine and terrestrial environments.

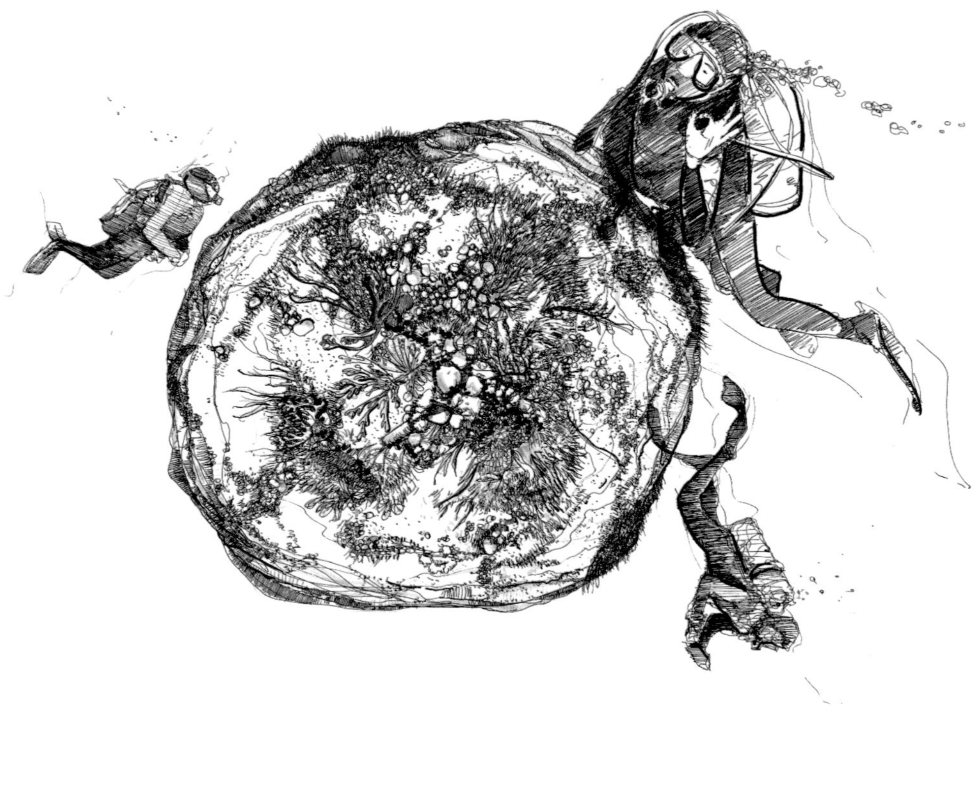

Figure 8.1 | Perception of underwater landscape immersion. Image by Fabien David/Atelier Coloco.

Exploring the underwater world

Immersion in the marine environment offers points of contact for all the senses in a state of gravitation, drifting, and loss of the horizon, all of which profoundly cause one to question their relationship with the landscape. Here, it is no longer a question of a face-to-face encounter between a gaze and an expanse, as on land. The imagination assembles scattered, disordered, disproportionate fragments of knowledge. The view is only a few meters away—even in clear waters—and the landscape becomes personal, intimate, and impossible to photograph in its entirety.

Representations of the underwater world have developed in parallel with the exploration of the depths and have long been limited to images of fantastic creatures. This environment remains largely unknown and difficult to access, with most dives only reaching depths of 40 meters. Beyond this depth, robotic technological equipment is often used. This 'frontier' continues to stimulate the imagination, which explains why underwater landscapes—although rarely visited—are very present in our daily lives via fantasized images. This invitation to the imagination of the landscape is an essential opportunity in our contemporary crisis of representations of a world in which increasing disturbances to our common ecosystem are occurring.

Worldwide, nearly 30% of the CO2 emissions produced by mankind are captured by the ocean (Centre National de la Recherche Scientifique, 2015). The ocean serves a major role in reducing the human impact of climate change, representing one of the most promising avenues for research into a more respectful relationship with marine ecosystems. In terms of regulating climatic phenomena, the benefits of marine environments are immense. Moreover, marine environments regenerate much more quickly than terrestrial environments. On a local scale, the threats to marine biodiversity are diverse. However, certain causes of coastal ecosystem degradation are particularly severe on the entire Mediterranean coastline. First, the anthropization of the coastline and the development of aquatic activities make the richness and good ecological state of the seabed a factor of attraction. Despite this, the anchoring of boats degrades the seabed and contributes to the dispersion of biological pollution. The impacts of metropolitan areas represent the main environmental disturbance, which directly or indirectly weakens the environment and degrades the underwater landscape.

Calanques National Park has a unique configuration since it is the only terrestrial and marine national park in an urban Mediterranean environment, covering more than 400 km^2. This territory is home to a diversity of ecological contexts that produce a wide variety of habitats—and thus a wide variety of landscapes—ranging from the shallowest, most easily accessible coastal areas to the most inaccessible canyon areas up to 3,000 meters deep. These situations are characterized by morphological data which are occupied by species adapted to each habitat, producing one of the most beautiful

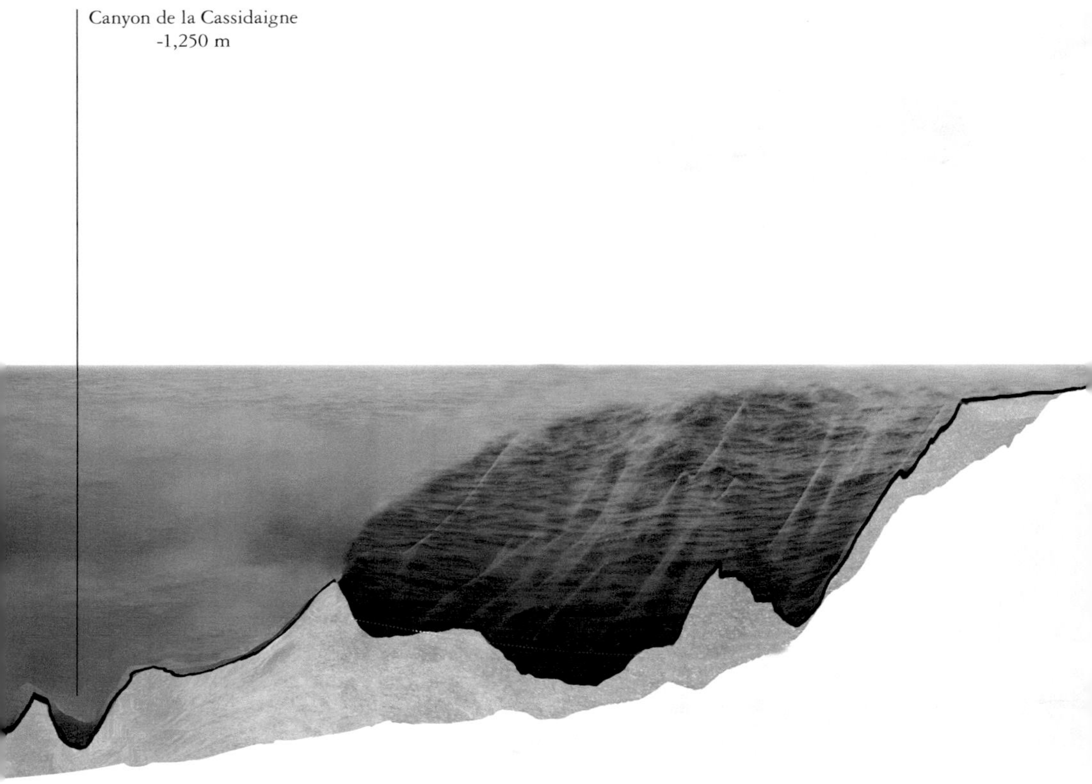

Figure 8.2 | Transverse section through littoral conditions. Image by Fabien David/Atelier Coloco.

Mont Carpiagne
646 m
Niveau 0 de la mer

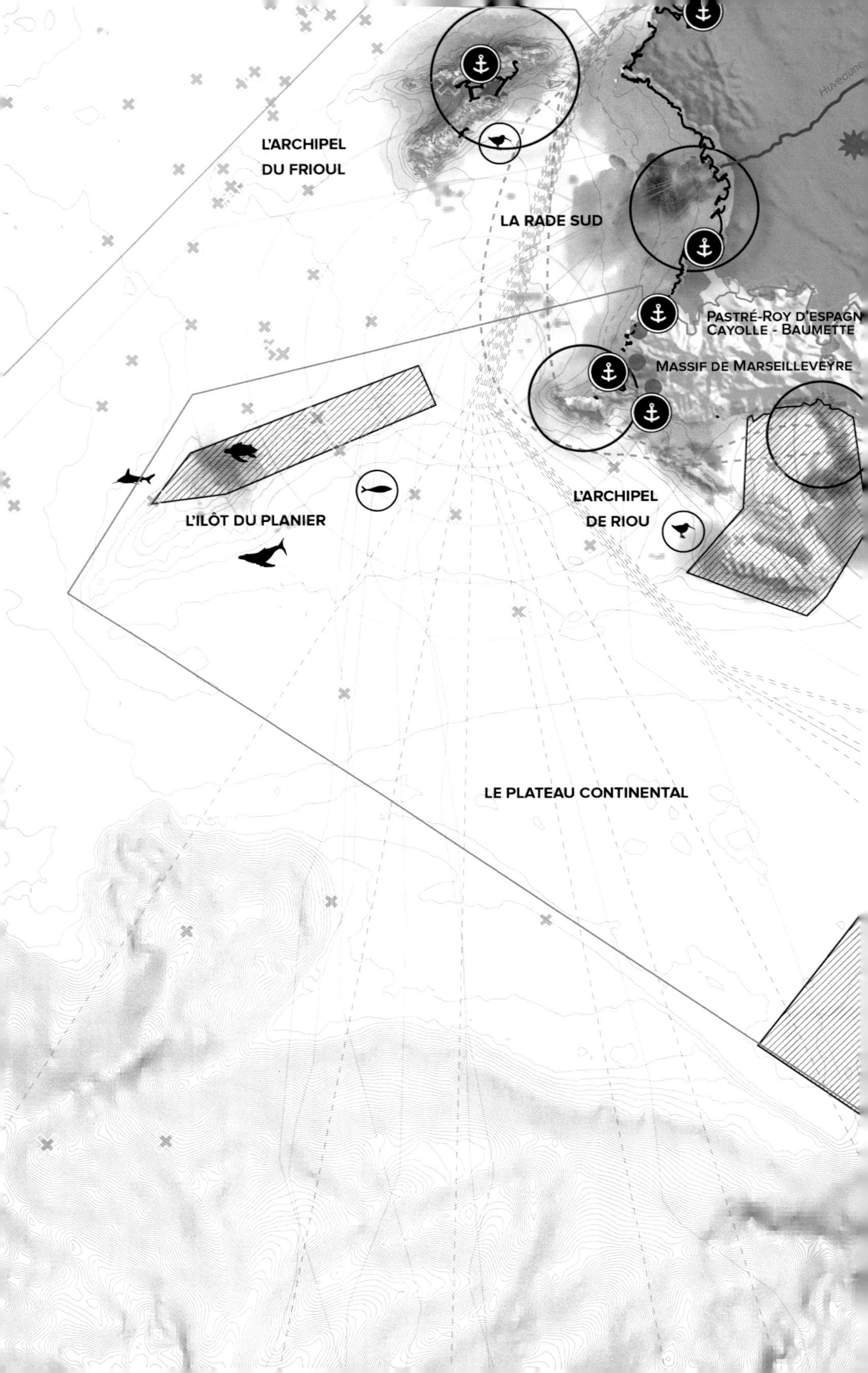

Huveaune
L'ARCHIPEL
DU FRIOUL
LA RADE SUD
PASTRÉ-ROY D'ESPAGN
CAYOLLE - BAUMETTE
MASSIF DE MARSEILLEVEYRE
L'ÎLOT DU PLANIER
L'ARCHIPEL
DE RIOU
LE PLATEAU CONTINENTAL

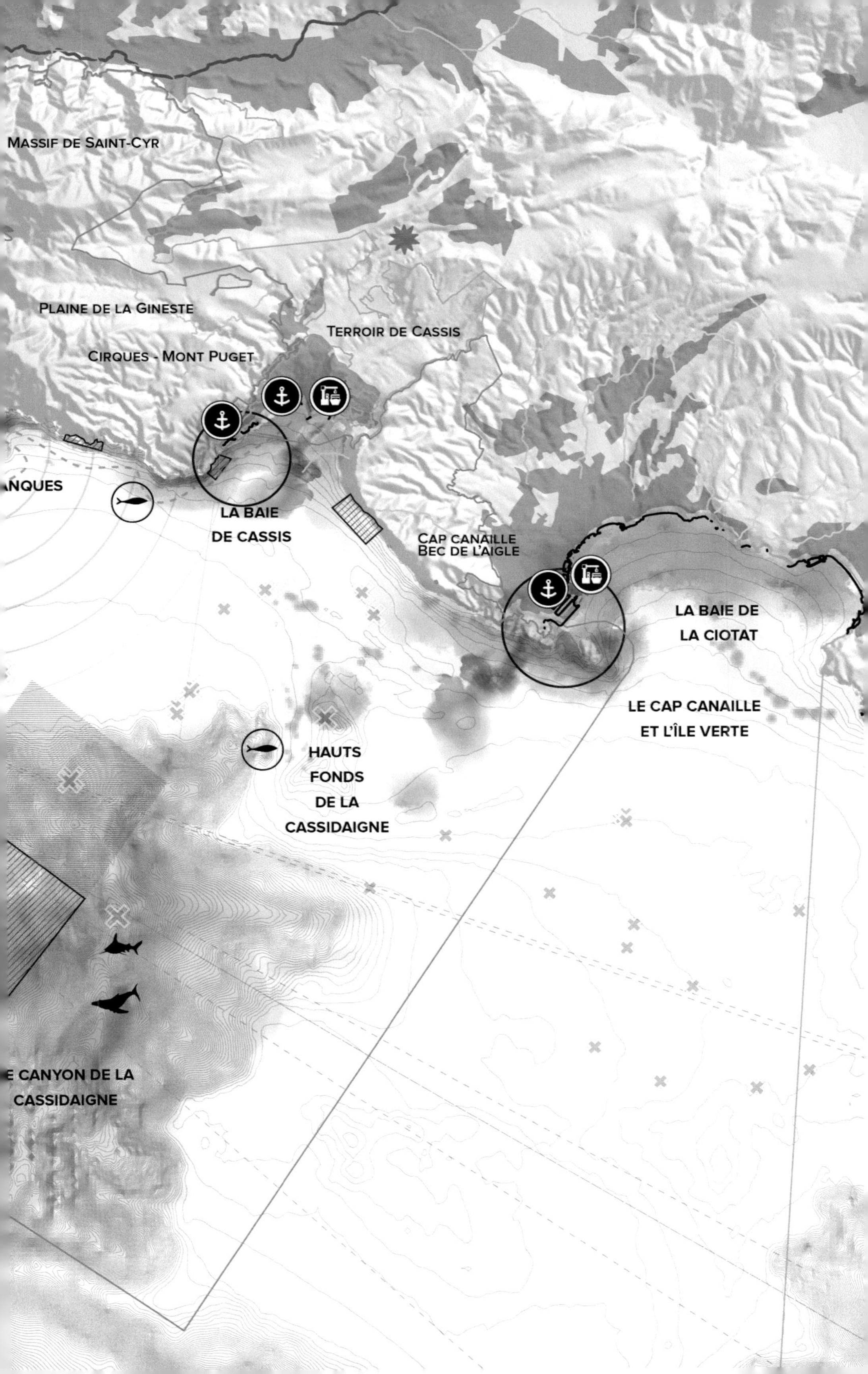

Massif de Saint-Cyr
Plaine de la Gineste
Terroir de Cassis
Cirques - Mont Puget
ANQUES
La Baie
de Cassis
Cap Canaille
Bec de l'Aigle
La Baie de
La Ciotat
Le Cap Canaille
et l'Île Verte
Hauts
Fonds
de la
Cassidaigne
E Canyon de la
Cassidaigne

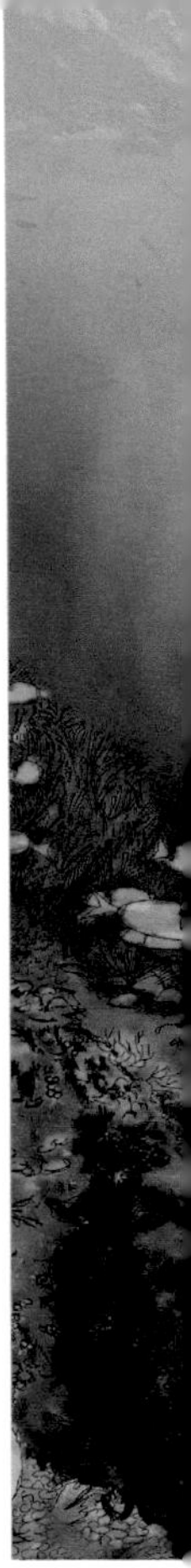

Previous page, Figure 8.3 | Map of pollution and anthropic impact on underwater landscapes. Image by Fabien David/Atelier Coloco.

Figure 8.4 | Landscape reconstitution by diving and drawing perceptions. Image by Margaux Bidat/Atelier Coloco.

landscape mosaics in the Mediterranean. However, at present, the reference nautical chart, which accumulates information very partially, without any representation of the underwater landscapes ignores all the diversity of life and landscapes and only integrates bathymetric data and a few landmarks for navigational safety. Underwater, nothing is more diverse than the constantly changing seabed, which is subject to the dynamics of currents and winds.

The creation of the National Park in 2011 confirms the public's commitment to combining nature conservation and accessibility for the metropolitan public. In the context of the current environmental emergency, the issue of preserving marine environments is becoming a major priority. This context raises the following question: How can we share the questions, research, and solutions that allow us to take care of our reservoir of oxygen, food, and fantastic dreams with the public and politicians?

The implementation of a territorial project through a Landscape Plan confirms the transdisciplinary role that the notion of landscape can bring, which has only been very recently extended to underwater spaces. Simultaneously considering the landscape as both a subject and a tool makes it possible to appreciate and enhance what often remains little known and rarely visited. This study proposes an approach to creating a common vision between actors in the terrestrial landscape, and, for the first time, the underwater landscape. Olivier Laroussinie, Director of the Marine Protected Areas Agency, stated the following: 'The philosophical and scientific stakes of underwater landscapes are extremely important because this concept will enable the crossing of different disciplinary fields, question accepted concepts (ecosystem, habitat, biocenosis, biotope, etc.), and offer new avenues of reflection in connection with new currents of thought and research objects.'

This experience represents a new type of collaboration in terms of methodology. The large team comprising members with different approaches shares the process along participative moments of ateliers. The proposal of (co)designing a project for new relationships with other species leads to intervention opportunities and principally integrates non-intervention. New perspectives on facing the climate crises will focus on biodiversity as a condition. The time required for adaptative evolution in underwater landscapes could be significantly reduced if human pressure is reduced while allowing free dynamics to move forward.

These situations are characterized by morphological data that are occupied by species adapted to each habitat, producing one of the most beautiful landscape mosaics in the Mediterranean. However, the reference nautical chart, which accumulates information very partially without any representation of the underwater landscapes, presently ignores all the diversity of life and landscapes while only integrating bathymetric data and a few landmarks for navigational safety. Underwater, nothing is more diverse than the constantly changing seabed, which is subject to the dynamics of currents and winds. In this common dimension, implementing a landscape plan involves the following aspects:

Understanding the evolution of landscapes over time and defining the framework for this evolution.

Guaranteeing the coherence of interventions on the territory in the long term for the protection of life.

Providing the means to build a concerted quality territorial project supported by the actors in all relevant fields (research, fishing, tourism, town planning, heritage, etc.).

In practice, this work took place over two years. It was based on dives at emblematic sites, a partnership approach to organizing thematic workshops, and coordination meetings to consider different perspectives according to the approaches and professions of both public and private participants. It brought together approximately 100 people, who pooled their experience and knowledge. This project feeds into research and action in the various fields of intervention at sea. It also involved formulating objectives that will be carried forward politically. This approach, unprecedented in its methodology and formalization, intends to serve as a tool for ecological protection through the recognition of landscapes, as well as the establishment of land-sea solidarity and a lever for the emergence of a common landscape culture across of the wider territory.

This initiative must be pursued by establishing a permanent working team to coordinate actions and assess impacts on underwater landscapes. The stakes are immense since our lives depend on the marine environments. Thus, we must act urgently to protect these environments, which are difficult to understand but critically important to preserve.

References

Centre National de la Recherche Scientifique (2015) *Ocean and Climate: Scientific Notes.* Available at: <https://www.ocean-climate.org/wp-content/uploads/2015/06/150601_ScientificNotes.pdf> [Accessed 20 January 2023].

Parc National des Calanques, Atelier Coloco paysagistes, Clément, G. and Onf, and Bosc, S. (2016) *Plan De Paysage Du Parc National Des Calanques.* Parc National des Calanque, Agence d'urbanisme de l'agglomération marseillaise, République Française, Direction régionale de l'environnement, de l'aménagement et du logement de Provence-Alpes-Côte d'Azur. Available at: <https://www.calanques-parcnational.fr/sites/calanques-parcnational.fr/files/atoms/files/pncalanques-plandepaysage-oct2016_2.pdf> [Accessed 29 November 2023].

Part II

Solid | Ice

9 | (Co)Designing With Time

A Tempo-Material Approach Along Arctic Island Coasts

Eimear Tynan

Design Project Data

Location
Bjørnøya (Bear Island) | Norway | Europe
Project Team
Eimear Tynan
Date
2017 | 2022
Coordinates
74° 50' 38.50" N | 19° 00' 13.83" E
Area
Selected coastal sites on the island (precise area not applicable)
Length
n.a.
Investment
n.a.

Bjørnøya Project

Landscape architects are increasingly confronted with environmental changes that go far beyond physical disruption and destruction, extending to social, political, economic, cultural, and ecological concerns. These changes are unfolding over different time frames—from disrupted seasonal trends to more long-lasting changes occurring over decades, centuries, and millennia. These scales of time can be difficult to comprehend because they often exceed human time and challenge us to imagine pasts and futures where more-than-human agencies and processes are dominant. To address such broad pluralities and time scales, tempo-material approaches are proposed that involve exploring time through close engagements with the materialities of a site. This approach is applied along the Arctic island coast of Bjørnøya, where a range of diverse coastal materials are explored to engage with changes that have occurred and are currently taking place over different frames and scales of time. This situated mode of research is a form of (co)designing that emphasizes temporalities concerning the states, conditions, and processes inherent to these materials. It is argued that this approach offers a means to work with time and environmental change, thus offering hope in confronting local and global concerns and crises.

Introduction

Often regarded as a peripheral region, the Arctic has taken center stage in discussions concerning the climate crisis. A report from the Arctic Monitoring and Assessment Programme (2021) states that the Arctic region is warming three times faster than the global average. The consequences are far-reaching and are not limited to the Arctic region. Some of the more concerning effects of a warming climate include the thawing of permafrost, which has the potential to release large volumes of greenhouse gases such as carbon dioxide and methane into the atmosphere. Additionally, a reduction in sea ice means that there is no longer a buffer to protect vulnerable coastal areas, thus hastening coastal erosion. Less sea ice also means that Arctic oceans are more accessible for shipping and oil and gas exploration and exploitation. Futhermore, the melting of glaciers has a concerning local impact with regard to the availability of drinking water and hydroelectric power. On a broader spatial and temporal scale, melting glaciers are expected to significantly raise sea levels globally. A final consequence to note is the steady migration of woody plant species into more northerly regions of the Arctic, which is transforming local ecologies and hydrologies over a wide area.

The list of environmental consequences linked to a warming Arctic is a long one. The changes precipitate through the Earth's atmosphere, biosphere, hydrosphere, and lithosphere, all of which occur over varying timeframes and timescales. These concerns raise pressing questions about hope for the present and future. The *tipping points* of climate change (or points of no return) that earth scientists often refer to have either been reached or nearly reached. Unsurprisingly, climate extremes and crises are increasingly causing anxiety. Perhaps this is because hope has been lost, or perhaps we no longer know what we are hoping for. After all, many of the changes being experienced are unpredictable, drastic, and challenge our preparedness for the future and the approaches of adaptive design measures. These have spurred the discipline of landscape architecture to consider and design for futures that recognize and accept the continually shifting nature of our landscapes across a range of different paces and durations of time.

As a result, landscape architecture has been forced to consider time differently from what was once adequate for the needs of the discipline. For example, the cyclical patterns of the seasons and the growth and distribution patterns of vegetation have been significantly disrupted in many regions of the world. The unrelenting rise of sea levels and flooding add to the list of pressing issues for the discipline. This has spurred many landscape architects to confront these unpredictable and extreme environmental changes and to design differently through anticipatory means. The subject of time invisibly infiltrates the environments we work with and the people who live in these environments. This chapter will shine a spotlight on time and how landscape architects can and should engage with it more creatively in confronting rapidly changing environments.

This study will focus on the coast of the Arctic island of Bjørnøya, a small island of 178 km^2 located in the Barents Sea at 74° 26'N, 19° 02'E (see Figure 9.1). The island is a part of the extensive Svalbard archipelago in the Norwegian Arctic. The northern part of the island is relatively flat and measures 15 km at its widest point. The terrain is mostly rocky and pock-marked with numerous small lakes. The southern tip of the island, lying 20 km from the north coast, is composed of jagged cliffs of up to 400 m in height. Despite the island having a rich history of visitors to its shores, including whalers, walrus hunters, fox hunters, scientists, and miners, it has no permanent population. The only inhabitants on the island are those working at the island's meteorological station.

For this study, Bjørnøya is of interest for two main reasons. First, there is a very rich mix of materials along its shores that display traces of former activity on the island. Second, the material changes and conditions on this island are worthy of study in light of a rapidly changing climate, with the island's local meteorological station records supporting this notion. Notably, the island and the surrounding sea are warming up. Sea ice, which once had a regular and predictable presence, has become unpredictable and less common. Sea bird populations are also changing on the island. It is expected that vegetation on the island will also change over time with the gradual establishment of woody species. All of these factors warrant a tempo-material approach to investigate some of the changes that are currently occurring on Bjørnøya.

This chapter proposes (co)designing with time to actively engage and understand material changes that are occurring along a selected Arctic coast. The first section, 'Hope Through (Co)Designing With Time,' introduces four different kinds of hope that characterize types of attitudes toward changing environments. Following this, I identify with one type of hope called 'authentic hope,' which aligns with my tempo-material approach to working with a coastal environment under environmental stress. In the second section, 'A Tempo-Material Approach,' I describe the relevance of this approach in the context of an Arctic coastal environment. The third section, 'Finding Time Along Bjørnøya's Coast,' activates and applies the tempo-material approach by utilizing methods commonly used in landscape architecture. The chapter draws to a close by emphasizing how (co)designing with time can offer a valuable and meaningful way of working with change, thus offering 'authentic hope' in a rapidly changing Arctic.

Hope Through (Co)Designing With Time

Before I proceed to describe and demonstrate ways to (co)design with time, I wish to position how this approach can offer hope in an era of rapidly

Figure 9.1 | A map showing the island of Bjørnøya, with its location in the inset provided. Image by Eimear Tynan.

changing living environments. A study published in 2020 by researchers Andrew Park, Elizabeth Williams, and Melanie Zurba analyzed and appraised the multifaceted nature of hope among conservation workers, educators, and others involved in environmental protection. In the study, they acknowledged that there are many variations of hope based on their review of a range of literature highlighting different types of hope. However, they managed to home in on four main categories: Authentic (or active), passive, false, and radical hope. According to the authors, a sense of agency and motivation to act is strongly associated with authentic hope, which may be achieved on an individual or collective level. Added to this is the motivation to establish a pathway to reach intended goals. Meanwhile, passive hope is where there is a generalized hope for a positive outcome but 'personal agency is either absent or not exercised' (Park, Williams and Zurba, 2020, p. 3). False hope is 'characterized by inappropriate (unattainable) goals, illusory future expectations, and poorly thought-out action strategies' (Park, Williams and Zurba, 2020, p. 6). Finally, the authors describe radical hope as post-apocalyptic in that it offers 'hope for future revival or redemption, albeit in a form that eludes current conception, and for which no current pathway exists' (Park, Williams and Zurba, 2020, p. 3).

These four categories of hope may be directly linked to a philosophical model proposed by landscape architect Dirk Sijmons (2021). The model, in the form of a quadrant, establishes four distinct philosophical positions for the discipline of landscape architecture in relation to the Anthropocene. The first position is called *denialism* and is characterized by either humans being uninformed about the Anthropocene and its direct relation to climate change, or denial of the existence of the Anthropocene. From this position, conservatism dominates and land is directly connected to property rights. Passive hope and false hope could be added to this position whereby action to improve environments is denied, particularly regarding long-term strategies and consideration for the more-than-human. The second position that Sijmons described is *modernist/ecomodernist*. This takes a rational perspective and places much faith in resolving problems through technology—even if the technology does not yet exist. Therefore, there is a form of optimism that is not necessarily accompanied by a roadmap. This position resonates strongly with Park's (Park, Williams and Zurba, 2020) notion of radical hope.

The third position posited by Sijmons is *posthumanism/ontological pluralism*, which brings forth ethical considerations that extend to humans and more-than-humans. Sijmons (2022) stated that this position is 'sympathetic to nomadism and akin to the light treading on the earth associated with indigenous peoples' (p. 120). It is a position that is very much rooted in the present while revolving around ideas of co-existence. In this case, the type of hope offered here could be described as 'authentic hope' because action and change are embedded in this philosophy. However, despite evidence of this position being goal-oriented, it is unclear whether roadmaps or ways to work harmoniously and actively into the future are a part of this philosophy.

The final position is called *Anthropocentrism 2.0* where there is a shift from humans being a dominant agency of change and destruction to a species that inherits a sense of responsibility and care. Sijmons (2021) claimed that this 'offers a gloomier but likely more realistic diagnosis of the situation than a posthumanist stance' (p. 123). Action is a strong feature of this position, which incorporates working *with* natural processes to resolve environmental issues. The type of hope offered by this position is firmly 'authentic hope,' whereby action is guided through informed and situated pathways.

Considering the four categorizations of hope by Park (Park, Williams and Zurba, 2020), and the four positions outlined by Sijmons, my research most closely aligns with the authentic hope positioned in *Anthropocentrism 2.0.* My 'authentic hope' comes into being from an individual level. I am broadly informed by climate research but work toward my own research goals to elicit new interpretations of time and materiality through a selection of Arctic island coasts.

A Tempo-Material Approach

This practice-driven approach offers an alternative way of thinking and knowing about Arctic island coasts through a time-centered perspective. It regards coasts as moving, dynamic spaces where materials move across and through the land, sea, and air. It emphasizes the role of time as materials move through coastal sites. These move at very different tempi and over broad timescales. Attention toward each material encourages consideration of its immediate local context as well as thinking about its journeys before it arrives at the site and after it leaves. Although the approach is specifically applied to a site, it entices the designer to think beyond local spatial and temporal scales. The approach is composed of practicing time in three distinct ways where the *time of the environment*, *time of the method*, and *time of the designer* are considered.

Time of the Environment

This gives attention to the materials present in a specific environment; in this case, an Arctic coastal site. The conditions and states of the materials, as well as the different processes actively playing out, are considered. How time is playing out along these coasts is directly linked to how different methods are deployed and what information they provide. An example of this is the use of audio recordings. In a coastal context, this could typically include the sound of the waves encountering the coastal zone with seabirds squawking in the background. From this, different temporalities can be detected through the materiality of the water, coastal materials (in the form of rock, sand, or pebbles), and birds. There is the cyclical process of the tides, the rhythmic sound of the waves, the deep time qualities of the sand, and the spontaneous sounds of the birds. Such recordings reveal busy, overlapping compositions of sounds that provide insights into the range of temporalities playing out in a particular environment.

Time of the Method

Following on from the previous point is attention toward the tools and techniques that may be used to examine the time and materiality of a site. Different tools and techniques operate over very different scales of time and have the capacity to draw out a range of different material temporalities. For example, using a camera to take one photograph captures a specific place at a specific moment in time. However, time-lapse photography that uses a sequence of photographs with precise and equal intervals of time allows for different information to be depicted that may not have been discernible for a person observing the same scene. For slower environmental phenomena, such as the build-up and movement of sea ice, time-lapse photography is a suitable technique. This will be elaborated further in the next section of the chapter.

Time of the Designer

This encompasses the experiential, affective, reflective, and intuitive responses of the designer to the coastal site. Time spent *in* an environment allows the designer to become conscious and aware of different conditions and processes affecting various materials. There is also time spent *away* from a site, which is also valuable since it allows the designer to process information and reflect on aspects relating to a site, as well as relevant fieldwork practices. The time of the designer is often expressed through field notes, sketches, or annotations on maps and drawings.

The time of the environment, time of the method, and *time of the designer* are the foundations of a tempo-material approach. Collectively, they offer opportunities to examine a diverse range of materials and consider the varying and overlapping notions of time that activate a particular environment. In the following section, I will demonstrate how to apply a tempo-material approach based on research conducted on the Arctic island coast of Bjørnøya.

Finding Time Along Bjørnøya's Coast

> 'Another freezing day on Bjørnøya, but the sun is shining and I am eager to get outside. I walk towards Kvalrossfjære. It feels magical revisiting this beach again, but this time everything is frozen solid. When I arrive

> at the beach, there are no signs of any old walrus bones, driftwood, or plastic that I had photographed previously. They must be buried under the snowdrifts this time. It is difficult to determine exactly where the land meets the sea, but some crinkled striations offer clues as to where the waves have come to a halt by the low temperatures. Globules of ice are strewn across the beach, peppered with long sun-dial shadows. While photographing, I notice a very peculiar sound coming from the sea that I have never heard before. It resembles a slow, high-pitched, creaking wooden door followed by a series of fast-paced thudding sounds. As I approach what I perceive as a firmly frozen sea, I realise it is the sea ice heaving to the moving tide. I spend the next 40 minutes or so sitting on large floats of sea ice contently recording these alien sounds.'
>
> (Bjørnøya, March 8, 2019)

My visit to Bjørnøya took place in March 2019, when there was an unexpected build-up of sea ice around the island. Working in freezing conditions meant that the methods I used were carried out as efficiently as possible. I was dependent on a range of tools to help me observe. In addition to keeping field notes during this visit, I deployed a range of different fieldwork methods, including handheld photography, time-lapse photography, drone photography, and audio recordings. After I returned to mainland Norway, I continued processing material gathered on site and produced maps, models (Tynan, 2021), and illustrations.

The following is a selection of three methods—time-lapse photography, audio recordings, and illustrations—that were integrally tied together as a working sequence to specifically explore sea ice and the different integrated time dimensions.

Time-Lapse Photography

Time-lapse photography is a method that can capture the flow and movement of material phenomena over a specific period. This method was chosen for two main reasons. First, it was convenient to set up a camera on a tripod to observe various changing phenomena while allowing me to conduct other work during my time on Bjørnøya's shores (see Figure 9.2). Second, this method was useful for documenting the work I was doing while recording the conditions around me. For example, time-lapse photography showed how I was conducting audio recordings and moving on the sea ice.

Time-lapse photography can show changes over a duration of time that differs from single-shot handheld photography. During fieldwork visits, it is not possible to observe and process all the events unfolding around the designer. This method helps in bridging this shortfall even though the camera is fixed to one viewpoint and cannot replace the 360-degree freedom

Figure 9.2 | A range of photographs over a three-hour period showing changing wind and light conditions at Bjørnøya's meteorological station in March 2019. Time-lapse photographs by Eimear Tynan.

that a human observer can afford. Despite this, it is a useful and flexible method used to record material change and movement.

Audio Recordings

Michael Fowler, a musician whose work involves the study of sound in architecture and landscape architecture, noted that using sound at the early stages of the design process acts as a 'mediating language between listener and environment' (Fowler, 2013, p. 113). In the Arctic, where the surfaces can dramatically and temporarily change from stone to snow or from liquid sea to ice, sound recordings are very effective in capturing these material changes. The changes conditioning the materials are brought about by processes beyond the site, such as winds, temperature changes, and tides. Since these processes and conditions are not adequately captured through photography, audio recordings supplement new readings of this coastal site.

As I previously pointed out in my field notes at the beginning of this section, the sounds from the sea ice were alien to me. The heaving tides below the sea ice resulted in high-pitched sounds originating between the compact islands of ice known as pancake ice. The audio file may be accessed from Tynan (2022). The subtleties of the tides, which were barely perceptible to the naked eye due to the frozen state of the sea, were amplified in the audio recordings. The recordings allowed for a fluidity of these materials to be evoked, which visual forms of representation such as photography and mapping cannot achieve.

Although audio recordings can offer minute and immersive details of a place, if one does not know where the sounds were recorded, they can sound placeless. Therefore, it was necessary to use audio recordings with other methods such as text, photographs, or illustrations for a more complete reading of these coastal environments.

Illustrating an Audio Recording

Concerning the final point I made in the previous paragraph, I translated the sea ice audio recording into an illustration that attempted to capture words describing the sounds and the movement of the sea ice with the tide. This method is experimental but it has the capacity for visual and audio representations to merge (see Figure 9.3). Words are the key ingredients to this graphic representation, and I initiate a foundation for an audio glossary to be developed further. The positioning of the words attempts to give a spatial reading of the ups and downs of the ice as it moves with the tides beneath it. This composition overlaps the sounds with movement. For example, as the ice moved up and down, there was a distinctive groaning sound; however, when the sea ice was more stable or horizontal, hissing and creaking sounds prevailed. Illustrating sound is an exercise that can emphasize the more delicate characteristics of a site that photography cannot capture.

The four methods described in this chapter—field notes, time-lapse photography, audio recordings, and illustration—offer diverse interpretations of time and materiality within one coastal site. It is an example of how using different methods to (co)design with time can show how the subjective and experiential sensibilities of the designer are deeply embedded in the chosen methods and the conditions of a site.

Conclusion

(Co)Designing with time can help to address notions of hope where there is an active and participatory role from the designer. In the book *Arts of Living on a Damaged Planet: Ghosts of the Anthropocene*, the authors state the following: 'Perhaps counterintuitively, slowing down to listen to the world—empirically and imaginatively at the same time—seems our only hope in a moment of crisis and urgency' (Tsing et al, 2017, p. 8). In many ways, my tempo-material approach attempts to apply this perspective whereby close observation is required in seeking out varied temporalities of this coast through engagement with its materials. An attunement toward the coastal materials required consideration of the processes acting upon the materials, an appraisal of the conditions of the site, and personal reflections on the differing states of the materials.

Working with coastal materials can operate on several temporal scales. The deep, geological time of a grain of sand is very much under the influence of winds, tides, and anthropogenic disturbances, all of which have their own durative properties. This implies that we can stretch our thinking beyond the local or site scale and look at other agencies that greatly affect the temporal and physical attributes of a material. Since landscape architecture is a future-oriented discipline, I can speculate on how the materialities of these coasts are likely to change and the environmental, climatic, and political threats that are likely to multiply and unfold. Rather than deny these changes, it has been more productive for me to accept that all of the materials I encounter are on a journey.

crackling
groaning
hissing
high-pitched groan
thump!
thump!
thump!
thump!

crackling
squeaking
abrasion
high-pitch
crackling

crackling
hissing
groan rising

high groan
wood-pecker beating
downward squeak
hisssssss

loud increasing groan
high-pitch
crackling

hissing
gargling

d
o
w
n
w
a
r
d
groan
hissssssssss
steady

clear crackle
slight groan,
hiss

growing gr
decreasing g
hiss

stops
soaks
gurgles
groaning rising to high-pitch
fades to crackling

deep groan
loud squeak
wood-pecker beats
elongates
squeaks

ground rising squeaks
crackling
hiss

hissing
clear and then muffled
compression

clear crackling
tension
loud
slow
decreasing groan
stops to a halt

hissing
increased groan
high-pitched
wheezing

drains
hisssssss

low pitch groan
decresing
soaking
hiss

crackling
background groan
hissssssss

Figure 9.3 | A graphic interpretation of the audio recording from the sea ice along Bjørnøya's shores. Image by Eimear Tynan.

References

Arctic Monitoring and Assessment Programme (2021) *Arctic Climate Change Update 2021: Key Trends and Impacts: Summary for Policy-makers*. Tromsø, Arctic Monitoring and Assessment Programme.

Fowler, M. (2013) 'Soundscape as a design strategy for landscape architectural praxis.' *Design Studies*, 34 (1), pp. 111–128.

Park, A., Williams, E. and Zurba, M. (2020) 'Understanding hope and what it means for the future of conservation.' *Biological Conservation*, 244, pp. 1–11.

Sijmons, D. (2021) 'In the Anthropocene, Site Matters in Four Ways.' In: Kahn, A. and Burns, C. eds. *Site Matters: Strategies for Uncertainty Through Planning and Design*. London, Routledge, pp. 110–130.

Tsing, A., Swanson, H., Gan, E. and Bubandt, N. eds. (2017) *Arts of Living on a Damaged Planet: Ghosts of the Anthropocene*. Minneapolis, The University of Minnesota Press.

Tynan, E. (2021) 'What time is this coast? Temporal encounters in the Arctic.' *Journal of Landscape Architecture*, 16 (2), pp. 10–25.

Tynan, E. (2022) *Bjørnøya Sea Ice I, 2019: Sound Cloud*. Available at: <https://soundcloud.com/user-921067170/bjornoya-seaiceiwav?utm_source=clipboard&utm_medium=text&utm_campaign=social_sharing> [Accessed 29 September 2022].

10 | Building Up Water

Artificial Glaciers in Ladakh

Carey Clouse

Design Project Data

Location
Ladakh | North India | Asia
Project Team
Leh Nutrition Project | Ice Stupa Team
Date
1980 | present
Coordinates
34° 7' 27.48" | 77° 28' 0.804"
Area
n.a.
Length
n.a.
Investment
n.a.

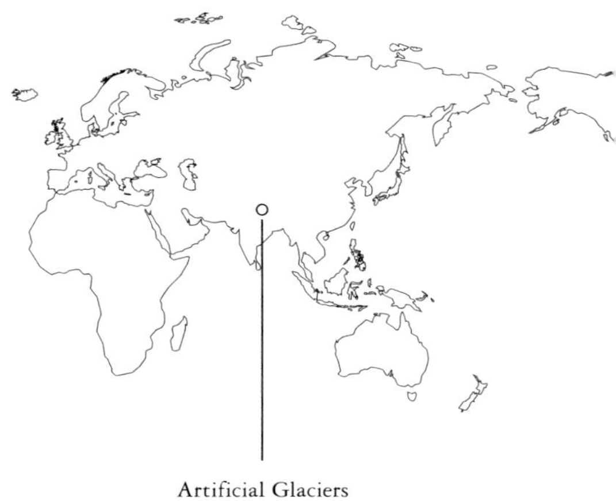

This chapter introduces the climate-adaptive artificial glacier projects of Ladakh, north India. These landscape ice masses have been produced in north India since the 1980s as a means of bolstering the region's stockpiled irrigation for use across seasons. In recent years, flat artificial glaciers have been adapted into other shapes, such as ice stupas—vertical pyramids of ice that also feature social, cultural, and religious aspects. These and other offshoots of artificial glacier engineering, such as icefall glaciers and automated ice stupas, highlight the wide variety of techniques and design concepts that can inform and even re-configure ice caches. Finally, this form of design thinking is presented as an optimistic, accessible self-help strategy in the face of climate change challenges, with multiple approaches and dimensions worth considering. Each of these examples has the potential to be scaled up with broader buy-in or translated to other regions. Artificial glaciers, icefall glaciers, and ice stupas each offer ways of integrating (co) design by people and natural forces in the service of adaptive environmental management.

Introduction

In desert climates, high-altitude meltwater is an extraordinarily valuable resource. Winter runoff from glaciers and snowfields is worth investing in by trapping, collecting, and saving it in stockpiles for agricultural use during the following growing season. In Ladakh, north India, a wide range of artificial glacier types offer methods for this form of water management. This chapter shares the history and design thinking of artificial glaciers, explores some of the current trends for (co)design by people and natural forces in this medium, and acknowledges the inherent optimism found in these types of Do It Yourself (DIY), self-sufficient climate change adaptation efforts.

In one form or another, artificial glacier projects have existed in Ladakh for more than three decades (Holden, 1998; Agarwal and Narain, 2000; Dawa, Dana and Namgyal, 2000). The earliest examples of this particular infrastructure involved forming a flat pool of ice with various diversion strategies explored through working prototypes that spanned decades (Norphel, 2012). However, artificial glaciers in Ladakh have become increasingly popular over the past decade due to the recognition of global climate change vulnerabilities as well as the promotion and dissemination efforts of the Ice Stupa Project (Chostak, 2016; Kolbert, 2019). Today, artificial glaciers are being adopted, adapted, and transformed by various user groups and in different regions, with these prototypes highlighting the creative and wide-ranging solutions available (Gladfelter and Yonetti, 2018; American Geophysical Union, 2022).

Artificial glaciers, as they have been conceived of in Ladakh, fundamentally appear as giant stockpiles of ice (see Figure 10.1). While distinct designs exist and have been piloted across the region, they might be broadly characterized as a repository of water held as ice in the winter for use in the following year (Than, 2012). Although this ice can exist in many different shapes, sizes, and scales, all designs share a standard engineering approach. Water is typically harvested in the winter, drawn from melting high-altitude glaciers and snowfields using gravity, diversion, and retention (Sharma and Bharat, 2017). In these designs, ice that is stored over the winter can then be tapped by farmers in the spring and summer, primarily for agricultural use.

Ladakh

Ladakh is a union territory and contested border region in north India, abutting Pakistan and China. This location, between the Karakoram and Himalayan Mountain ranges, has long been activated by its proximity to major international trade routes (Rivzi) and has also become a destination for tourism since 1974. Despite major international influences and their associated economic activity, Ladakh has also been routinely characterized as a remote place with small, self-sufficient villages practicing subsistence agriculture (Norberg-Hodge, 1991). The Indian government has intervened in visible ways over the past several decades by building new roads, schools, hydro and solar power projects, wells, and public toilets, and by offering residents a regular ration of subsidized staples (Demenge, 2010; Dame and Nüsser, 2011; Dame et al, 2019). Although these inputs have helped to bring modern services and amenities such as electricity, Wi-Fi, plumbing, education, medical care, and transportation access to the region, such interventions also frequently fail to deliver the promised resources (Demenge, 2013). The Ladakhi landscape is now perhaps best read as a complex system of interdependencies between locals, visitors, and the State, despite the long legacy of villages operating as self-supporting, independent units (Aggarwal, 1997).

Despite numerous transformations to Ladakhi lifestyles through the region's experience with globalization, the physical makeup and operation of small villages remain relatively unchanged. Due to Ladakh's mountainous terrain, many villages have been opportunistically sited between high-altitude glaciers and lower-lying riverbeds. These subsistence farming enclaves are composed of dozens of households, proximate to walled-in fields. Winter snowfall helps to build up glaciers in the region, which provide meltwater to lower-lying villages. Moreover, any unused surface flow proceeds beyond the village borders to feed the headwaters of the Indus River.

Many of Ladakh's villages have been self-sufficient entities for hundreds if not more than a thousand years (Osmaston, 1994). Their reliance upon meltwater, originating from high snowfields and glaciers, is readable on the landscapes since villages located in the creased valleys of massive mountains stand out from the dry desert context through their verdant walled footprints (see Figure 10.2). As water resources shift from year to year, this legible outline may shift in response; however, it always serves as a reminder that agriculture is only possible where water can be trapped, diverted, held, and used to irrigate crops. In conjunction with building the soil, this relationship with water is a central tenet of land husbandry in Ladakh and one that has been developed and refined into an elegant and equitable process over many years (Gutschow, 1993; Crook and Osmaston, 1994; Gutschow and Mankelow, 2001; Sharma, 2019).

Climate Change in Ladakh

Climate change has impacted the relationship between farming villages and water husbandry as meltwater has become less reliable (Grossman, 2015; Chostak, 2016). Since Ladakh is located in the rain shadow of the Himalayan mountains, between the Karakoram mountains, it receives only 115 cm of annual rainfall (Thayyen et al, 2013). Climate change-induced precipitation patterns have changed over the years, threatening to disrupt 'the delicate balance between the historical climate and ecology of the region' (Chevuturi, Dimri and Thayyen, 2018, p. 544). A study of the 121 small glaciers and 60 cirque and valley glaciers in the Trans-Himalayan Kang Yatze Massif

Figure 10.1 | An Ice Stupa in Ladakh.
Photograph by Nishant Tiku.

Figure 10.2 | The lush, cultivated fields in Zanskar stand out against a desert backdrop. Photograph by Carey Clouse.

reported a 14% retreat from 1969 to 2010 (Schmidt and Nüsser, 2017). However, the authors of that study noted that such a picture of widespread retreat is complicated by the variability of each glacier and season, thereby making an argument for considering the impacts of glacier erosion on a case-by-case basis, rather than at a regional level.

In addition to the shrinking natural glaciers above Ladakh, a changing climate can also be seen in new types of weather in the region (Yangkey, 2020). Glacial lake outburst floods (GLOFs) occur when a frozen glacier melts and then bursts through mud or ice dams, flooding downstream landscapes. This phenomenon has been documented for centuries in Ladakh (Sangode et al, 2011), but is rapidly becoming a cryosphere hazard consideration for regional risk analysis (Schmidt et al, 2020). Cloudbursts have also impacted Ladakh, inundating landscapes with intense bouts of rainfall. In 2010, the cloudburst near Leh caused mudslides and unstable slopes, killing 234 people and destroying parts or all of some 1,450 households (Gupta, Khanna and Majumdar, 2012; Ziegler et al, 2016). These major weather events appear to be the new normal in Ladakh, occurring with greater frequency and intensity. In addition to these acts, everyday regional weather patterns also appear to be shifting. Broadly, precipitation has been decoupled from seasonal norms (Yangkey, 2020). Water that may have fallen during freezing months may now come in warmer months, bringing rain instead of snow. While less pronounced and relatively poorly recorded, these changes can combine to transform glaciated landscapes.

For Ladakhi villages, a dependence on glaciers and snowfields for gravity-fed crop irrigation has required pivoting or adapting to this changing cryosphere. While high-altitude desert farming has always been a challenging undertaking, the reduction of regular, reliable irrigation water causes additional hardship (Angchok, Stobdan and Singh, 2014). This somewhat experimental work is largely DIY, self-driven, and falls into the category of adaptation (rather than mitigation), thus reflecting the desire of farming households to preserve a way of life that has been sustainable for centuries. Today, under the additional pressures of climate change, farming in Ladakh requires flexibility, creativity, and entrepreneurial thinking (Nüsser and Baghel, 2016).

Many villages make agricultural decisions as a collective. Thus, ensuring buy-in from the entire group presents an additional challenge to any climate-adaptive project (Sharma, 2019). Some villages have responded to the increased drought simply by reducing the acreage of farming parcels (Dame and Mankelow, 2010). This contraction can be detrimental to families who require larger croplands to sustain their lives and livelihoods, but it can also be viewed as a benefit for families who would otherwise outsource agricultural labor because family members are away at school, serving in the army, or have found work in the tourism industry as hotel workers, guides, or drivers (Dame et al, 2019). Other households may limit the number of livestock they keep because pasturing and winter feeding might not be feasible in drought conditions. Finally, others remain actively engaged in climate-adaptive efforts, from shoring up irrigation channels and ponds (to reduce shrinkage) to diverting water from new sources, and even adding new climate-adaptive designs, such as snow barrier bands or artificial glaciers (Nüsser, Schmidt and Dame, 2012; Grossman, 2015).

Artificial Glaciers

The use of artificial glaciers in Ladakh represents just one way that village groups have decided to address water scarcity under climate change. Each site yields a different ice mass for harvesting, and the physical forms of these built-up caches shift across the arc of the year as the glaciers grow and then diminish. Engineers have always used the term 'artificial glacier' as a shorthand way of evoking the idea of the large, upstream parent glaciers that have historically fed—and continue to feed—Ladakhi croplands. However, employing this nomenclature for artificial projects may be somewhat misleading because such production occurs at a far smaller scale and with limited enduring transference.

History of Artificial Glaciers

Artificial glaciers began appearing in Ladakh in the 1980s as the brainchild of Chewang Norphel, Ladakh's first 'Glacier Man' (Vince, 2009). Chewang Norphel was an engineer working with local NGOs at the time, searching for ways to manage irrigation systems more judiciously. At that time, farmers relied almost exclusively on glacial meltwater for a thin yet reliable source of irrigation, and the original artificial glaciers designed and built by Chewang Norphel and the Leh Nutrition Project enabled better husbandry. Chewang Norphel has now retired, having designed more than a dozen artificial glacier projects in Ladakh. NGOs such as the Leh Nutrition Project continue to carry on this work using many of the same physical strategies envisioned by Chewang Norphel.

In the winter of 2013/2014, the flat artificial glacier concept went through a major shift in both engineering and characterization. The Ice Stupa Project, led by Sonam Wangchuk, reimagined not only the diversion channels feeding artificial glaciers but also the shape and volume of these ice masses. The resulting towers of ice are called ice stupas, named for their striking resemblance to the Buddhist masonry chortens that dot the landscapes of the region (see Figure 10.3). The Ice Stupa Team took an agricultural irrigation strategy and promoted it as something more: A climate change adaptation project and a celebration of regional Buddhist identity—a sacred offering and ecological endeavor (Gladfelter, 2018). The

Figure 10.3 | The mountainous landscape is echoed in the white stone stupas in Ladakh. Photograph by Klara Avsenik.

Ladakhi ice stupa has garnered considerable attention in both national and international news media. While these stockpiles of ice may be built by communities or individual farmers for their own practical use, they have also emerged as landscape spectacles, serving as sites for tourism, cultural events, and collective prayer (see Figure 10.4).

The ice stupa effectively ushered the work of artificial glaciers into a new era, moving design conceptualization beyond pragmatic, utilitarian irrigation schemes into a more coherent multi-functional landscape intervention (Clouse, 2019). Indeed, the entangling of ecological, environmental, agricultural, social, and religious goals into a single landscape feature is a new way of conceiving climate adaptation efforts that carries promise for future work. Climate-adaptive projects could gain traction by appealing to more than environmental topics through the integration of multiple groups of stakeholders and additional visibility through widespread access and use.

The popularity of the ice stupa has led to new experiments and interventions in other parts of the world. For instance, a glaciologist at the University of Fribourg in Switzerland, Suryanarayanan Balasubramanian, has explored the mechanization of the ice stupa as a means of demonstrating broader applicability, utility, and efficiency (American Geophysical Union, 2022). Ice stupas have also been piloted in Nepal as a means of reducing GLOF hazards to downstream communities. Icefall glaciers, which pair the vertical orientation of the ice stupa with the landform armature of the artificial glacier, have also been built in Ladakh.

(Co)Design of Artificial Glaciers

The emergence of artificial glacier strategies in Ladakh over the past three decades offers an example of climate-adaptive design for the future. Here, design thinking has driven a host of new solutions to water scarcity that use affordable, accessible, DIY approaches. Each piloted project has been seen as a work in progress—experiments in an ever-evolving effort to fine-tune engineering. New ways of conceptualizing this work through the addition of religious, social, and cultural meanings, have given new dimensionality to what was originally conceived of as a straightforward irrigation device.

This work is inherently optimistic by addressing an otherwise oppressive and out-of-scale climate change phenomenon with a design that can be made with human hands. Even in the mountain landscapes of Ladakh, these interventions stand out as significant landforms, such as a series of cascading ice pools over a mile in length in the case of Phuksey Phu, or ice stupa towers that crest 100' in height (Kumar-Rao, 2020). However, the philosophy undergirding these projects remains human-scaled. Each of these climate adaptation strategies foregrounds the needs of individual farming households, who work together in the community by taking matters into their own hands.

Figure 10.4 | The ice stupa serves as a site for engagement as well as political, environmental, social, cultural, and religious uses. Photograph by Nishant Tiku.

DIY Design Adaptation

Part of the central ethos of each artificial glacier project is to create something useful with minimal tools, effort, and imported products. Both Sonam Wangchuk and Chewang Norphel sought to reduce the number and type of imported goods required to create these ice stockpiles. Gravity moves water through channels or pipes, and sites are opportunistically selected based on elevation, slope, and solar aspects. New developments in automation may make artificial glaciers more accessible while simultaneously threatening the affordability and DIY nature of this work.

Design Thinking for the Future

The artificial glaciers that were pioneered in Ladakh in the 1980s continue to develop and transform. Sonam Wangchuck has postulated that these types of ice caches could be useful in the Alps and the Andes, where water and ice are also in decline, or in Nepal to pre-empt GLOFs. Similarly, Chewang Norphel has theorized that his design for artificial glacier pools could easily be adopted by other dry high-mountain agricultural environments or communities. Notably, Chile and Kyrgyzstan have already produced ice stupas (American Geophysical Union, 2022).

Ladakh's intentional ice creations are part of a broader portfolio of design explorations that offer power in diversity. Climate change adaptation efforts cannot be static and they cannot all be the same; we will need to harness many different solutions to the climate crisis. Furthermore, we will need affordable solutions that can be scaled up, as well as contextual responses for uncommon conditions. It is unlikely that the answer to climate change will ever be a single, straightforward fix. Such answers will require the creativity of designers, who can make landscapes work again by working in their home environments. Perhaps the latent optimism of artificial glaciers comes from their plucky existence: These are giant marvels created by stakeholders working with limited materials, funding, and technology. These irrigation projects are audacious in their very existence, and they set the groundwork for other types of climate change adaptation work. They were made by people with limited training in design attending to an out-of-scale environmental problem and willing to try something new in the face of climate change. Beyond any single design scheme, it is the approach of these individuals that provides a model for creative future practice.

Acknowledgments

Many thanks to Nishant Tiku, Chewang Norphel, the leadership and staff of the Leh Nutrition Project, Sonam Wangchuk, and the rest of the Ice Stupa Team for sharing their thoughts about artificial glaciers with me. I owe a debt of gratitude to the National Geographic Explorer Foundation as well as the Fulbright-Nehru organization for sponsoring my fieldwork in Ladakh.

References

Agarwal, A. and Narain, S. (2000) 'Water harvesting: community-led natural resource management.' *ILEIA Newsletter*, 16 (1), pp. 11–13.

Aggarwal, R. (1997) 'From Utopia to Heterotopia: Towards an anthropology of Ladakh.' In: Osmaston, H. and Nawang, T. eds. *Recent Research on Ladakh 6: Proceedings of the 6th International Colloquium on Ladakh, Leh, 1993*. First Asian Edition. Delhi, Motilal Banarsidass Publishers, pp. 21–28.

American Geophysical Union (2022) *Automation Builds Bigger, Better Ice Tower Reservoirs for High, Dry Farming*. Available at: <https://phys.org/news/2022-06-automation-bigger-ice-tower-reservoirs.html> [Accessed 1 August 2022].

Angchok, D., Stobdan, T. and Singh, S.B. (2014) 'Community-based irrigation water management in Ladakh: A high altitude cold arid region.' *Twelfth Biennial Conference of the International Association for the Study of Commons, Governing Shared Resources: Connecting Local Experience to Global Challenges*, pp. 1–12.

Chevuturi, A., Dimri, A.P., and Thayyen, R.J. (2018) 'Climate change over Leh (Ladakh), India.' *Theoretical and Applied Climatology*, 131 (1–2), pp. 531–545.

Chostak, S. (2016) 'Local adaptation strategies to climate change: Learning from Ladakh.' *ILI Law Review*, winter, pp. 7–24.

Clouse, C. (2019) 'Climate-adaptive design: Building up Ladakh's ice stupas.' *Landscape Journal*, 38 (1–2), pp. 25–41.

Crook, J. and Osmaston, H. (1994) *Himalayan Buddhist Villages: Environment, Resources, Society and Religious Life in Zangskar, Ladakh*. Bristol, University of Bristol.

Dame, J., Schmidt, S., Müller, J. and Nüsser, M. (2019) 'Urbanisation and socio-ecological challenges in high mountain towns: Insights from Leh (Ladakh), India.' *Landscape and Urban Planning*, 189, pp. 189–199.

Dame, J. and Mankelow, J.S. (2010) 'Stongde revisited: land-use change in Central Zangskar.' *ERDKUNDE*, 64 (4), pp. 355–370.

Dame, J. and Nüsser, M. (2011) 'Food security in high mountain regions: agricultural production and the impact of food subsidies in Ladakh, Northern India.' *Food Security*, 3 (2), pp. 179–194.

Dawa, S., Dana, D. and Namgyal, P. (2000) 'Water harvesting technologies and management system in a micro-watershed in Ladakh, India.' In: Chalise, S. and Banskota, M. eds. *Waters of Life-perspectives of Water Harvesting in the Hindu Kush-Himalayas: Volume II*. Kathmandu, ICIMOD, pp. 235–259.

Demenge, J. (2010) 'Road construction, dependency, and exploitation in Ladakh, North India.' *Journal of Workplace Rights*, 15 (3–4), pp. 303–326.

Demenge, J. (2013) 'The road to Lingshed: Manufactured isolation and experienced mobility in Ladakh.' *HIMALAYA, The Journal of the Association for Nepal and Himalayan Studies*, 32 (1), pp. 51–60.

Gladfelter, S. (2018) *Ladakh's Artificial Glaciers, Ice Stupas, and Human-made Ice Reserves*. Fulbright-Nehru Student Research Report. United States-India Educational Foundation, p. 31.

Gladfelter, S. and Yonetti, E. (2018) 'Ladakh's artificial glaciers, ice stupas, and other attempts to survive a warming planet.' *Tibet Himalaya Initiative*, 11 March. Available at: <https://www.colorado.edu/tibethimalayainitiative/2018/03/11/ladakhs-artificial-glaciers-ice-stupas-and-other-attempts-survive-warming-planet> [Accessed 27 December 2019].

Grossman, D. (2015) 'As Himalayan Glaciers Melt, Two Towns Face the Fallout.' *Yale Environment360*. Available at: <https://e360.yale.edu/features/as_himalayan_glaciers_melt_two_towns_face_the_fallout> [Accessed 2 August 2022].

Gupta, P., Khanna, A. and Majumdar, S. (2012) 'Disaster management in flash floods in Leh (Ladakh): A case study.' *Indian Journal of Community Medicine*, 37 (3), p. 185.

Gutschow, K. (1993) 'Lords of the fort, lords of the earth, and no lords at all: politics of irrigation in three Tibetan societies.' In: Osmaston, H. and Nawang, T. eds. *Recent Research on Ladakh 6: Proceedings of the 6th International Colloquium on Ladakh, Leh, 1993*. Leh, Ladakh, pp. 105–115.

Gutschow, K. and Mankelow, S. (2001) 'Dry winters, dry summers: water shortages in Zangskar.' *International Association for Ladakh Studies*, 15, pp. 28–32.

Holden, C. (1998) 'Artificial glaciers to help farmers.' *Science*, 282 (5389), pp. 619–619.

Kolbert, E. (2019) 'The ice stupas.' *The New Yorker*, 20 May, p. 54.

Kumar-Rao, A. (2020) 'One way to fight climate change: Make your own glaciers.' *National Geographic Magazine*, July, pp. 98–108.

Norberg-Hodge, H. (1991) *Ancient Futures: Learning from Ladakh*. San Francisco, Sierra Club Books.

Norphel, C. (2012) 'Artificial glacier: A high altitude cold desert water conservation technique.' In: *In Defense of Liberty Conference Proceedings*. New Delhi, India, Leh Nutrition Project, p. 3. Available at: <http://indefenceofliberty.org/story/4343/4495/Abstracts-of-papers--Session-3--April-9-10Climate-Change-Understanding-Himalayan-Ecology> [Accessed 24 November 2013].

Nüsser, M. and Baghel, R. (2016) 'Local knowledge and global concerns: Artificial glaciers as a focus of environmental knowledge and development interventions.' In: Meusburger, P., Freytag, T. and Suarsana, L. eds. *Ethnic and Cultural Dimensions of Knowledge*. Cham, Springer International Publishing (Knowledge and Space), pp. 191–209.

Nüsser, M., Schmidt, S. and Dame, J. (2012) 'Irrigation and development in the Upper Indus Basin.' *Mountain Research & Development,* 32 (1), pp. 51–61.

Osmaston, H. (1994) 'The farming system.' In: Crook, J.H. and Osmaston, H. eds. *Himalayan Buddhist Villages: Environment, Resources, Society and Religious Life in Zangskar, Ladakh.* Bristol, University of Bristol, pp. 139–198.

Sangode, S.J., Phadtare, N.R., Meshram, D.C., Rawat, S. and Suresh, N. (2011) 'A record of lake outburst in the Indus valley of Ladakh Himalaya, India.' *Current Science (Bangalore)*, 100, pp. 1712–1718.

Schmidt, S., Nüsser, M., Baghel, R. and Dame, J. (2020) 'Cryosphere hazards in Ladakh: the 2014 Gya glacial lake outburst flood and its implications for risk assessment.' *Natural Hazards*, 104 (3), pp. 2071–2095.

Schmidt, S. and Nüsser, M. (2017) 'Changes of high altitude glaciers in the Trans-Himalaya of Ladakh over the past five decades (1969-2016).' *Geosciences; Basel*, 7 (2), p. 27.

Sharma, A. (2019) 'Giving water its place: artificial glaciers and the politics of place in a high-altitude himalayan village.' *Water Alternative*, 12 (3), pp. 993–1016.

Sharma, A. and Bharat, K. (2017) 'One's waste, another's right: translating history and making the Ladakhi commons.' In: *IACS Conference Proceedings*, pp. 1–28. Available at: <https://www.iasc2017.org/wp-content/uploads/2017/06/11N_Arjun-Sharma.pdf> [Accessed 2 August 2022]

Than, K. (2012) 'Artificial glaciers water crops in Indian Highlands.' *National Geographic*. Available at: <http://news.nationalgeographic.com/news/2012/02/120214-artificial-glaciers-water-crops-in-indian-highlands/?rptregcta=reg_free_np&rptregcampaign=20131016_rw_membership_r1p_us_se_w#finished> [Accessed 24 November 2013].

Thayyen, R.J., Dimri, A.P., Kumar, P. and Agnihotri, G. (2013) 'Study of cloudburst and flash floods around Leh, India, during August 4–6, 2010.' *Natural Hazards*, 65 (3), pp. 2175–2204.

Vince, G. (2009) 'Glacier man.' *Science*, 326 (5953), pp. 659–661.

Yangkey, T. (2020) *An Analysis of Precipitation Trend in Leh, Ladakh, Northern India*. Durham, Duke University. Available at: <https://sites.duke.edu/gcstenzinyangkey/files/2020/04/Tenzin_GCSthesis.pdf> [Accessed 2 August 2022].

Ziegler, A.D., Cantarero, S.I., Wasson, R.J., Srivastava, P., Spalzin, S., Chow, W.T.L. and Gillen, J. (2016) 'A clear and present danger: Ladakh's increasing vulnerability to flash floods and debris flows.' *Hydrological Processes*, 30 (22), pp. 4214–4223.

11 | The Sounds of Melting Landscapes

Facing the Climate Crisis with Sonic Intimacy in the Alps

Ludwig Berger

Design Project Data

Location
Morteratsch Glacier | Switzerland | Europe
Project Team
Ludwig Berger | Dennis Häusler | Johannes Rebsamen | Matthias Vollmer | Chair of Landscape Architecture Christophe Girot
Date
2015 | 2018
Coordinates
46° 25' 2.168" N | 9° 55' 52.161" E
Area
n.a.
Length
n.a.
Investment
n.a.

Melting Landscapes

The course series and audiovisual project *Melting Landscapes* collectively documented a glacier in the Swiss Alps over several years. A group of researchers and students attached underwater microphones to the glacier to make the complex inner sound world of the disappearing ice audible. The recordings were presented in installations, at concerts, and as an audiovisual publication. This chapter contextualizes the project and reflects on the reactions of listeners. It explores the ways in which the sound of continuous melting on a small scale offers different temporal and spatial perspectives on the ongoing climate crisis, and how listening to glaciers may enable an intimate, embodied knowledge of the climate crisis.

The Sounds of Melting Landscapes

The Alps are a hotspot of the climate crisis in Europe in two respects. First, the Alpine countries are responsible for a large share of global CO2 emissions. For example, the small country of Switzerland has an indirect influence on at least 2–3% of global CO2 emissions (Wenger et al, 2022). Second, temperatures in the Alps are increasing approximately twice as fast as the global average. Glaciers are particularly affected by this heating; they are predicted to lose more than 80% of their current mass by 2100 (Hock et al, 2019). Can listening to alpine glaciers help people grasp the reality of these facts and figures? If so, what change of perspective on the climate crisis can be stimulated by sonic media?

From 2015 to 2018, the project *Melting Landscapes* was developed by the Chair of Landscape Architecture Christophe Girot at ETH Zurich.[1] To witness the rapidly retreating glaciers in the Alps, and thus the effects of climate change on our own doorstep, we visited the Morteratsch Glacier in the southeastern part of Switzerland (Upper Engadin) in different seasons with several groups of students. Equipped with various microphones as well as large-format analog cameras, we took acoustic and visual samples of the retreating glacier tongue. In winter, we pressed underwater microphones to the ice using snow as an adhesive mass. Once the microphones made contact with the glacier ice, they acted like stethoscopes on a body, revealing the complex sounds occurring in the glacier.

In the summer, we placed hydrophones into the glacier puddles. In this process, sounds that were not perceivable from the outside became audible with bare ears: melodic squeaking, clicking and rattling, gargling and gurgling, hissing and fizzing, as well as deep droning. Numerous sounds from microscopic melting processes were repeatedly interrupted by large ice blocks breaking off. What looks like a solid mass from the outside sounds like a living body from the inside.

We have made our recordings available to the public as a vinyl record with a photo book,[2] as spatial sound installations,[3] as a concert series in collaboration with musicians,[4] and in sonic lectures. Like the glacier itself, the project takes many different forms. It was collaboratively designed by the glacier, the various recordists and their instruments, the artists, and the listeners (who interpret the results).

The project opens sensory access to glaciers and the climate crisis for listeners, thus creating the potential for 'intimacy without proximity' (Metcalf, 2008). Listeners expressed associations with hearing a human body or hearing sounds in the womb. They developed an appreciation for—and even empathy and kinship with—a previously elusive phenomenon. The bodily experience of the melting process, with all its liveliness, facilitates the development of a new perspective on glaciers. Images comparing existing glaciers to their past are well known, much like data on ice loss

Figure 11.1 | Ludwig Berger recording the glacier tongue in the winter of 2016. Photograph by Johannes Rebsamen.

Figure 11.2 | Students under the glacier tongue, recording the glacier in different entry points using hydrophones and piezo microphones and writing recording notes. Photograph by Ludwig Berger.

and predictions of future ice loss. However, sound offers a different temporal perspective: Not the before and after, but the continuous time in which the glacier is melting. This may offer something against a possible sense of irretrievability and fatalism. The countless delicate micro-events that reach our ears may teach us that it is not a matter of all or nothing, but that every intervention that can stop global warming is important.

These sounds also provide access to another spatial scale. On a visual level, one mostly perceives the events of a glacier on a large scale, like the collapse of ice blocks. However, in sound, one can experience the small scale from which the climate catastrophe takes its course. Every change in the temperature of a glacier is immediately audible as a sound because of increased melting activity. These microscopic sounds have the potential to remind us of the all-encompassing impact of capitalist societies in the natural world. Recording and collectively listening to the sounds of melting ice may enable an intimate, embodied knowledge of our crisis, which is the condition for adequately responding to it. The (co)design of *Melting Landscapes* unites the living activity of the glacier with the cultural lives of humans, providing hope through awareness of our commonality, interconnectedness, and responsibility.

Notes

1 | The project was co-organized by Dennis Häusler, Johannes Rebsamen, and Matthias Vollmer.

2 | See https://landscapearchitecture.bandcamp.com/album/melting-landscapes

3 | CLIMATE CARE | Vienna Biennale for Change 2021 (MAK Museum Vienna, Austria), Spazio Materia (Prato, Italy) 2021, Cima Norma Art Festival (Dangio, Switzerland) 2022, etc.

4 | Together with the Chuchchepati Orchestra (Patrick Kessler, Dieb13, Julian Sartorius).

Acknowledgments

Student recordings: Christina Akesson, Michael Beerli, Shalin Bhatt, Philipp Bosshart, Lloyd Broda, Altair Cerda Tirado, Weilun Chen, Jenyben Dave, Devashree Dvivedi, Camino Escosa García, Nadia Gaiarin, Masahiro Gokita, Andrea Gonzalez Palos, Camilla Gormsen, Benjamin Graber, Dennis Häusler, Yufei He, Anna Hermann, James Horkulak, Marie Jacobsen, Seonju Kim, Aleksander Kongshaug, Bess Laaring, Jungwoo Lee, Xijie Ma, Viktoriya Maleva, Stephan Mauser, Marc Over, Vasiliki Papadimitriou, Bhoomi Parmar, Shailaja Patel, Thalia Poziou, Timon Ritscher, Laura Schütz, Dhruvil Soni, Jean-Marc Stadelmann, Dharal Surelia, Tulsi Vadalia, Oliver Walter, Jan Westerheide, Sonja Widmer.

References

Hock, R., Rasul, G., Adler, C., Cáceres, B., Gruber, S., Hirabayashi, Y., Jackson, M., Kääb, A. Kang, S., Kutuzov, S., Milner, A., Molau, U., Morin, S., Orlove, B. and Steltzer, H. (2019) 'High Mountain Areas.' In: Pörtner, H.O., Roberts, D.C., Masson-Delmotte, V., Zhai, P., Tignor, M., Poloczanska, E., Mintenbeck, K., Alegría, A., Nicolai, M., Okem, A., Petzold, J., Rama, B. and Weyer, N.M. eds. *IPCC Special Report on the Ocean and Cryosphere in a Changing Climate.* Cambridge, UK and New York, Cambridge University Press, pp. 131–202.

Metcalf, J. (2008) 'Intimacy without proximity: Encountering grizzlies as a companion species.' *Environmental Philosophy*, 5 (2), pp. 99–128.

Wenger, F., Ziegler, M., Wulkop, A. and Keberle, A. (2022) *Klimastandort Schweiz, Schweizer Unternehmen als Globale Treiber für Netto-Null.* New York, McKinsey & Company.

Part III

Gas | Fog and Clouds

12 | The Cloud Gardeners

Water for a Thirsty Planet

Laura Cipriani and Gilles Clément

Design Project Data

1 | Fog x Beach
Location
Jamaica Pond | Boston | USA | North America
Coordinates
42° 19' 3.63" N | 71° 7' 13.082" O

2 | Fog Garden
Location
Chaumont-sur-Loire | France | Europe
Coordinates
47° 28' 41.336" N | 1° 11' 25.405" E

3 | Blur Building
Location
Yverdon-les-Bains | Switzerland | Europe
Coordinates
46° 46' 42.856" N | 6° 38' 27.296" E

4 | Nebolosus
Location
Zurich | Switzerland | Europe
Coordinates
47° 22' 0.897" N | 8° 32' 23.81" E

5 | La Geria
Location
Lanzarote | Canary Island | Spain | Europe
Coordinates
28° 59' 35.268" N | 13° 40' 9.07" W

6 | Namib Desert
Location
Namibia | Africa
Coordinates
24° 54' 34.212" S | 15° 19' 55.242" E

7 | Atacama Desert
Location
Chile | South America
Coordinates
24° 33' 22.69" S | 69° 25' 21.532" O

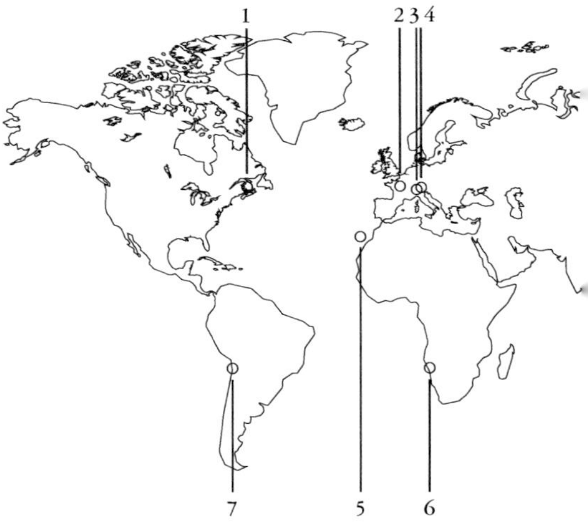

Some anthropic and natural practices involve working with the atmospheric phenomena of change of state, including clouds, fog, frost, and dew. This contribution explains how human and non-human 'cloud gardener' (Clément, 2021) practices can allow us to explore new (co)design techniques and open our imagination to new possibilities.

The introductory paragraphs 'Oceans of clouds. Drops of Fog,' 'The Rain Dance,' and 'Learning From Nature' by Laura Cipriani introduce readers to this theme by addressing the scientific-technical and poetic-aesthetic dualism of climate modification and adaptation by technologies, humans, plants, insects, bacteria, and microorganisms that oscillate between the two inverse phenomena of the passage of state: condensation and vaporization.

The last paragraph 'Clouds' is an excerpt from Clément's book *Nuages* (Clément, 2021) initially published in 2005. Clément explores clouds in the water cycle within the 'planetary garden' (Clément, 1999), intending to explain how cloud gardeners can 'create' and 'hold' water, frost, and fog in desert environments. From Chinese water towers to the Namib Desert, from the Atacama Desert to the everyday landscape of Clément in Limousin, France, cultivating clouds in dry regions of the planet also passes through the wisdom of gardeners who decide that their lawns should turn yellow. Letting plants organize with the new climate can provide hope in tackling the climate crisis.

Oceans of Clouds. Drops of Fog

Water is always unstable since it constantly moves in the water cycle. From the soil to the groundwater. From rivers to the ocean. From the sea to the atmosphere. Evaporation, transpiration, condensation, precipitation, infiltration, surface runoff, and subsurface flow are the physical processes of this constant movement. Today, climate change upsets water cycles and evaporation phenomena, increasing extreme weather events, droughts, and floods. Controlling the climate, or 'creating' rain, has always been considered a divine, non-human work. In many latitudes, the rain dance was a ceremonial rite to bring about precipitation while strengthening identity and social cohesion in the community (Merton, 1968; Chaves, 2010). The dance synthesized a (co)design of hope for an entire community to induce or 'cultivate' rain in certain areas of the planet.

This contribution describes working practices involving clouds, fog, frost, and dew. The intent is to talk about the gardeners of the clouds; that is, humans, animals, and plants capable of 'cultivating' water in conditions of scarcity and adapting to often extreme and inhospitable climates. Designing—or rather (co)designing—together with atmospheric phenomena is certainly not *the* solution for the changing climate. However, it allows us to open our imagination to new possibilities.

While we know that the global water balance at the planetary level remains essentially constant, the atmospheric phenomena of state change can fulfill temporary water needs. (Co)design through the condensation and vaporization of water—two inverse phenomena of the change of state—has given rise to a dualism of scientific-technical and aesthetic-poetic approaches to climate modification and adaptation by technologies, people, plants, insects, bacteria, and microorganisms.

The Rain Dance

Science has always tried to understand atmospheric mechanisms. Lamarck carried out the first scientific studies on clouds, which laid the foundations for modern meteorology in 1805 (Lamarck, 1805; Clemént, 2021). Using precise nomenclature, clouds are classified into 12 types based on their shape and altitude. Depending on their appearance, they are divided into isolated clouds with vertical development and extended clouds, elongated in layers. Based on their altitude in the sky, they are called *stratus* if closer to the ground or *cirrus* if further away. Even fog, composed of minute droplets of liquid water, is a ground-level cloud formed due to the condensation of water vapor in the atmosphere. It is precisely the phenomenon of condensation that involves a change in the state of matter from gaseous to liquid through heat removal.

From the middle of the previous century, technical sciences experimented with the possibility of interfering with clouds to encourage precipitation for the first time. This so-called *cloud seeding* is tested through the dispersion of certain substances—including silver iodide, dry ice, or salt—using planes, cannons, or ground devices to induce precipitation from the clouds (Laaksonen and Malila, 2022). In the 1940s, the initial discoveries of Schaefer (1946) and Vonnegut (1947) gave rise to intense research on 'cloud seeding,' initially in the United States. There are many reasons for using such technology: to limit enemy attacks during war events, to make up for a shortage of water in a specific region, to prevent hail on crops, to limit air pollution in particular circumstances (fires and radioactive episodes), and to increase visibility at airports during foggy conditions. In China, the most recent and ambitious study conducted to understand the mechanisms underlying the phenomenon of precipitation on a global scale has produced a planetary map of what are called *sky rivers*: 'water vapor flowing in the sky rivers is the atmospheric water that is most likely to precipitate onto the earth's surface' (Wang et al, 2018). To these studies, we add the interest in climate manipulation, which, in China, has given rise to the Chinese National Weather Modification Development Plan 2014–2020, in which the potential areas of experimentation are defined (Hunchuck, Ferrari and Cheng, 2021). Although such technology remains in an experimental phase, being able to interfere with, design, and modify the climate opens worrying scenarios of climate geopolitics, if we can call it that.

On the one hand, the engineering sciences are developing techniques and technologies to induce precipitation. On the other hand, landscape design has mainly tried to interact with the atmospheric elements with ephemeral, poetic, and aesthetic purposes by referring to vaporization (i.e., the change of state from liquid to gaseous). There are numerous climatic fog and steam installations. Japanese artist Fujiko Nakaya has been designing fog sculptures in many gardens worldwide since the 1970s (Theurer, Lissoni and Duguet, 2023), and one of her latest creations stands next to Olmsted's Emerald Necklace. Artistic practices, architecture, and the art of gardens have given rise to various suggestions that, although fascinating and helpful in stimulating debate and the imagination, may be limited to the scenographic aspect of working with the vaporous state. Just think of the temporary garden in Chaumont-sur-Loire in France by Peter Latz (Latz+Partners, no date), the famous Diller Scofidio machine for the '2002 Swiss Expo' in Switzerland (Diller Scofidio+Renfro, no date), the recent Nebolosus installation by architects Authos and Stella Speziali in the ancient botanical garden of Zurich (Authos, no date), the surrealist illusions of Bernard Smilde (see Chapter 1), or the Cloudscapes by Transsolar and Tetsuo Kondo Architects at the Venice Biennale (see Chapter 14).

In some desert areas of the world, *cloud farming* based on traditional knowledge and centuries-old practices for water condensation has proven promising in interacting with the vaporous state to harvest water. Fog nets

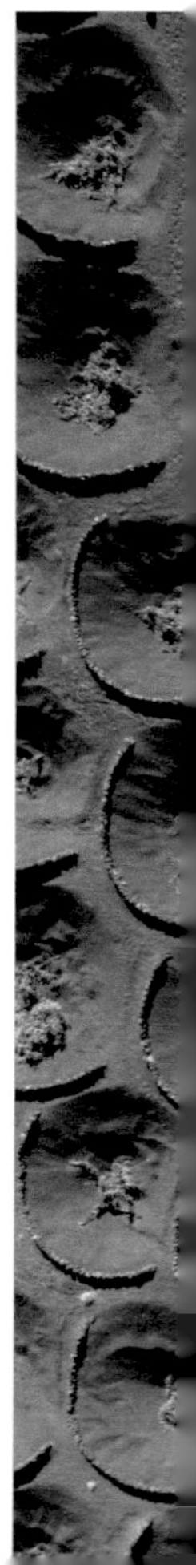

Figure 12.1 | The semicircular lava craters of Lanzarote in the area called La Geria are a traditional hydro-geological system used to collect water for wine production. Photograph by iStock.com/Donhype.

Figure 12.2 | Aerial view of the semicircular lava craters of Lanzarote. Photograph by iStock.com/Hector Martinez.

such as those of Peter Trautwein or Arturo Vittori (see Chapter 1) have optimized and reinterpreted a traditional system called *atrapaniebla*, which exploits the condensation of water in fog in Peru and the Atacama Desert of Chile through a collection system. Such technological devices not only satisfy the water needs of small, isolated communities and activate local agriculture and reforestation in rural areas but also trigger the participation and involvement of local communities in such projects (Correggiari et al, 2017).

Similarly, the ancient wine-related agricultural practices in the lava landscape of Lanzarote in Spain in an area called La Geria involve an ingenious traditional hydro-geological system that continues to allow the production of wine grapes and other crops in a semi-desert climate even today. The lunar volcanic landscape combines the aesthetics of a gigantic land artwork with the skillful reuse of water condensation. In the myriad of semicircular craters lie vines protected from the wind thanks to dry stone walls and covered for a few meters by fragments of volcanic rock—the so-called *picón*. The layer of picón captures humidity, rain, and dew during the night, preventing evaporation and releasing it during the day, thus ensuring correct nourishment for the plant. These volcanic sandboxes emerged after the eruption of Timanfaya volcano in the mid-18th century and are essentially based on water condensation into dew to alleviate the lack of precipitation in the area. Dew is simply the deposition of water droplets that form on cold surfaces with temperatures lower than the dew point due to the condensation of water vapor in the air. Therefore, the picón layer has multiple characteristics: it is capable of capturing water directly from the environment since it has vacuoles that facilitate the passage of condensed water (dew); it prevents evaporation since the iron layer insulates the soil, preventing moisture from escaping through evaporation into the atmosphere; it acts as an insulating mattress for air temperature since it captures solar radiation during the day and does not dissipate heat from the ground during the night (Morales, Ojeda and Torres, 2015).

All these examples show how humans—communities, scientists, artists, and architects—have tried different ways to (co)design with the atmospheric phenomena of condensation and vaporization. A water dance to capture clouds, fog, frost, and dew in which plant and animal species also actively contribute.

Learning from Nature

In some areas of the planet, mainly where precipitation is rare, plants, insects, bacteria, and microorganisms use water from the vapor state as a vital source of hydration. The Namib and Atacama deserts are extraordinary examples for understanding the mechanisms of plant and animal adaptation to extreme environmental conditions since multiple species have evolved to actively or passively collect water from fog or water vapor condensation (Gurera and Bhushan, 2020; Bhushan, 2020). This type of adaptation opens up new possibilities and potential experiments in times of climate change.

In the Namib Desert, a long and narrow coastal strip in southern West Africa, the unpredictable phenomenon of fog sustains the fauna and flora. Nearly 50 animal species use misty water in the liquid and vapor phases (Mitchell et al, 2020). Two species of beetles, *Onymacris unguicularis* and *Onymacris bicolor*, actively collect fog by basking in the mist (Hamilton and Seely, 1976) and capturing tiny droplets of water through the grooves of their carapace, finally conveying them to their mouths (Henschel and Seely, 2008). Other opportunistic species, although not active in collecting, drink the water from the fog deposited on vegetation or other surfaces. Moreover, beetles of the genus *Lepidochora* make structures to capture fog (Hamilton and Seely, 1976), while species such as *Onymacris unguicularis* emerge from their underground habitats to access misty water. To overcome unpredictable and episodic foggy events, some animals are even equipped with long-term internal water storage mechanisms.

Even plant species such as *Stipagrostis sabulicola*, a grass endemic to the central area of the Namib Desert, rely predominantly on fog for their water supply. The superficial structures of their leaves convey the collected water toward the base of the plant, allowing the efficient and rapid absorption of the atomized water by their superficial roots (Ebner, Miranda and Roth-Nebelsick, 2011; Roth-Nebelsick et al, 2012).

Even the Atacama Desert in Chile, one of the driest places on the planet, hosts multiple forms of plant and microbial life and is, by its nature, a laboratory for understanding the mechanisms of survival in extreme conditions. Although it was thought there was no life in such places, microorganisms have been found even in the most inhospitable areas. The fog advancing over the Chilean Coastal Range appears to be the main—if not the only—source of water for the microbial communities under the quartz of these hills (Azua-Bustos, Urrejola and Vicuñ, 2012). Although the minimum quantity of water necessary for microbial life in this environment and whether the microorganisms of the Atacama have evolved to be less dependent on water or have adapted remain unknown, it is evident that life persists with minimal quantities of water.

Desert plants also naturally evolved to thrive in arid, nutrient-poor conditions. Adaptive strategies common or specific to major Atacama plant lineages include the enrichment of bacteria that promote plant growth near their roots (Eshel et al, 2021). By sampling climate, soil, plants, and soil-microbial diversity, it is understood that associations between plants and microorganisms are essential for environmental survival. The soil close to the root system is enriched with bacteria that promote the plants' growth. Some cactus species, such as *Copiapoa haseltoniana*, native to the Atacama

Desert, use night fog as a supplementary source of water (Gurera and Bhushan, 2020). The epidermal surface collects fog and dew and is reflective when dry. The surface helps to increase water use efficiency by reducing the temperature difference between the plant and the air (Mooney, Gulmon and Weisser, 1977). Some cactus species are even equipped with conical spines with different integrated parts that serve various roles in the fog collection process depending on their surface structural characteristics (Ju et al, 2012).

To survive, plants exploit the humidity of the Pacific air, which is transported to the desert areas through fog. The Atacama Desert is also called a flowery desert since it transforms into a polychromatic and flourishing garden in peculiar and exceptional conditions. Under warmer temperature conditions due to El Niño currents, there is greater evaporation and increased precipitation in this area. When a few millimeters of rain and adequate temperature combine, they awaken seeds that have remained dormant for years. The desert is colored from lilac to an intense fuchsia thanks to a blanket of *pata de guanaco* (*Cistanthe sp.*) that covers the entire plain. With a progressive increase in rainfall, other blooms can be added: yellow and orange *añañucas* (*Zephyranthes bagnaldi*), white field *suspiros* (*Nolana baccata*), *malvilla* (*Cristaria sp.*) and *fraile coronilla* (*Encelia canescens*). It is a phenomenon as transient as it is poetic, which repeats itself with an unpredictable recurrence approximately every five to seven years and generates a complete ecosystem of insects, reptiles, and birds where even the fauna—which has also adapted to arid conditions—plays a crucial role in pollination and biodiversity. Desert plants arise from dormant seeds composed of an embryo, nourished by starch, and covered externally by a tough rind capable of spending five, seven, or even ten years in a latent state. In the last 40 years, about 15 events have occurred in the Atacama Desert.

The design of the landscape starts from a (co)design of different factors: plant and animal species have evolved to (co)design the landscape with fog and frost—so much so that the lessons of nature seem to surpass the ingenuity of humans in their beauty and wisdom. This is also the thesis Gilles Clément supports in the following chapter, taken from the book *Nuages* (Clément, 2021). The author explores clouds in the water cycle within what he calls the 'planetary garden' (Clément, 1999), which is the earth. Clément explains how cloud gardeners can 'create' and 'hold' water, frost, and fog in desert environments. Cultivating clouds in the dry regions of the planet also passes through the wisdom of the gardener, who lets plants organize with the new climate. The self-determinacy of plants provides hope in tackling the climate crisis. The human gardener can step aside and marvel at Mother Nature's gardeners' wisdom and renewing power.

Figure 12.3 | In the Namib Desert, the unpredictable phenomenon of fog allows the life of fauna and flora. Photograph by iStock.com/2630ben.

Figure 12.4 | The flowery desert of the Atacama manifests when a few millimeters of rain and adequate temperature combine. Photograph by iStock.com/Abriendomundo.

Clouds

Cloud: An aggregate of suspended microdroplets. Droplets are the result of condensation around a particle. Without particles—without 'impurities'—condensation does not occur. The particulate material covers a vast area, from waste dust to microorganisms. A cloud should not be considered water in vapor or dew but rather a complex, impure, and unformed whole.

The nature of the information enclosed in a cloud pertains to the nature of the impurities that make it exist. The wake of a jet aircraft is a cloud 'road' generated by the condensation of water vapor around kerosene particles. Transformed into rain, this cloud would be equivalent to an herbicide for a gardener. Other clouds loaded with nourishment act as fertilizer. A cloud combines the nutrients and the diluent necessary for assimilating that part of plants. We could feed a plant by making it pass through a cloud. The genus *Usnea*, 'beard lichen,' believed to live on the air of the time, feeds on tropical mists and low-lying clouds. Epiphytic plants care for every support: Tree branches, wires, electrical cables, pylons... A Usnea curtain, a wall of clouds, and the intimate meteorology of the bromeliads translate into an air saturated with water.

(...) A man jokes about the clouds, and the clouds protect him. Without them, the earth's temperature would be -18 °C on the ground, an ice desert. Their roof works like the glass of a gardener's greenhouse: Part of the solar radiation is trapped, maintaining an average temperature of 15 °C on the planet. This temperature depends on the natural composition of the atmosphere, in particular of the water vapor and future clouds. Gases such as CO2 (0.03% by volume) also minimally maintain the so-called greenhouse effect. Above all, the generalized concern about global warming comes from measurements revealing the real growth of this gas in the atmosphere and how this has contributed to an average temperature increase of 0.5 °C over the last century.

If the radiative balance remains balanced and livable for most species, the increase in temperature is accompanied by a series of climatic changes. Although all are known and analyzed according to more or less likely scenarios, they are always catastrophic for man.

(...) In certain countries, you have to go looking for (the clouds), interrupt their travel, modify their structure, and force them to turn into rain without bombing them, infiltrating them with poison, or altering the climate—without declaring war. Such gardening pertains to dry regions of the planet, where the weather conditions forbid all precipitation.

The Chinese have long cultivated clouds in cold deserts by building water traps as towers. The superimposed stones, neither grouted nor covered, form a massive concave building in the center. The humidity carried by night breezes passes through the building and condenses on the cold parts. Thus, the water drips on the sloping stones toward a well, where it collects. This device captures the little water from low clouds, such as in Namibia and the Atacama Desert, where the collection is more considerable.

In these districts where it never rains, plants and animals live. They go looking for water or obtain it for themselves. In Moon Plain, located in the Namibian desert between Windhoek and the coast, lives a beautiful plant called *Welwitschia mirabilis* for this very reason. Two low-lying generative assemblies each develop a striped leaf that follows continuous growth without any vegetative arrest. The desert wind tears the only two leaves into more petite strips. The oryxes take shreds of it. At the armpit of a stump, which becomes woody over the years, the brown and silky cones of the male and female apparatuses emerge. The tangled whole suggests an abandoned job, an unfinished chignon, a pile, a 'mop' … The first botanist who identified and recognized the rare qualities of this endemic species of Namibia, capable of growing in a sterile environment and without water, is told that finding it incredible, he fell on his knees before it. The *Welwitschia* collects nocturnal water through a gutter process. The thick ribs, oriented toward the heart of the plant, channel the water on the seats. The plant can live for 2,000 years or more.

A giant black beetle that lives in the sand does the same thing. During the day, it is buried under the dune. In the evening, it climbs a ridge, positions itself against the wind, and straightens its abdomen. Water condenses on its carapace, flows into the gutters formed by the edges of the elytra, and heads toward the oral apparatus.

None of this would exist without the presence of water-laden clouds that circulate at night, vanishing in the morning in the heat of the desert. Namibia's water machine works in a close relationship with the nearby ocean. During the day, warm desert air blows toward the ocean, condensing above a sea cooled by the Benguela Current from Antarctica. This freshness is accentuated by an ascent of deep waters under the constant influence of winds and currents, which attract plankton and fish to the surface (upwelling). At night, the north wind from the sea pushes the clouds over the desert without turning them into rain. However, to take advantage of them, beings exist that are capable of capturing them.

The same phenomenon occurs along the Atacama Desert, from Chile to the Peruvian coast. The cold Humboldt Current heads north, curves, and 'drowns' toward the equator after washing through the Galapagos archipelago. The permanent upwelling only disappears in the presence of El Niño. The rest of the time, clouds or night mists similarly cross the desert from west to east. In Chile, these cloud formations collide with the relief of the Andes, where they manage to flow into some snow at high altitudes. Elsewhere, the desert remains dry. On the edge of this desert and cold ocean, the people of Chugungo have no water. The pipelines from the big cities have long since given way. The deserted village dies. Not far away, the pre-

Figure 12.5 | Clouds.
Photograph by Tom Barrett.

Figure 12.6 | On the left, cloud classification according to Lamarck (Lamarck, 1805). On the right, names and types of clouds in the troposphere. Image by Gilles Clement and edited by Laura Cipriani.

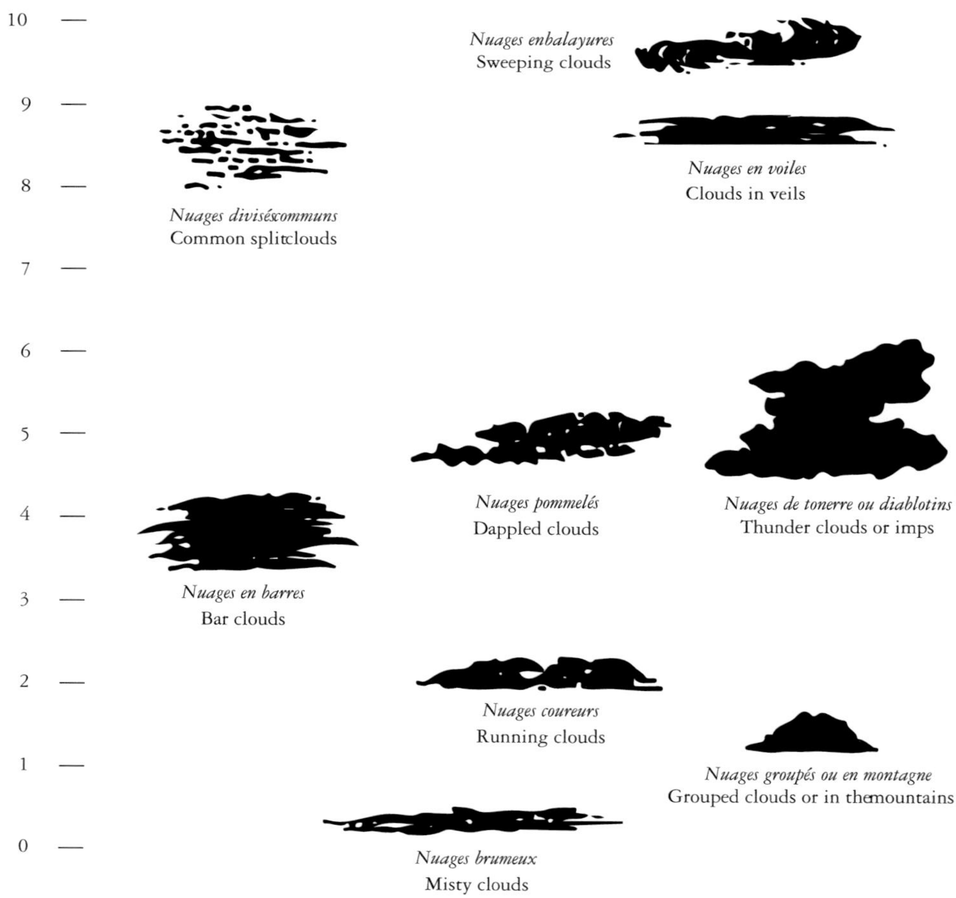

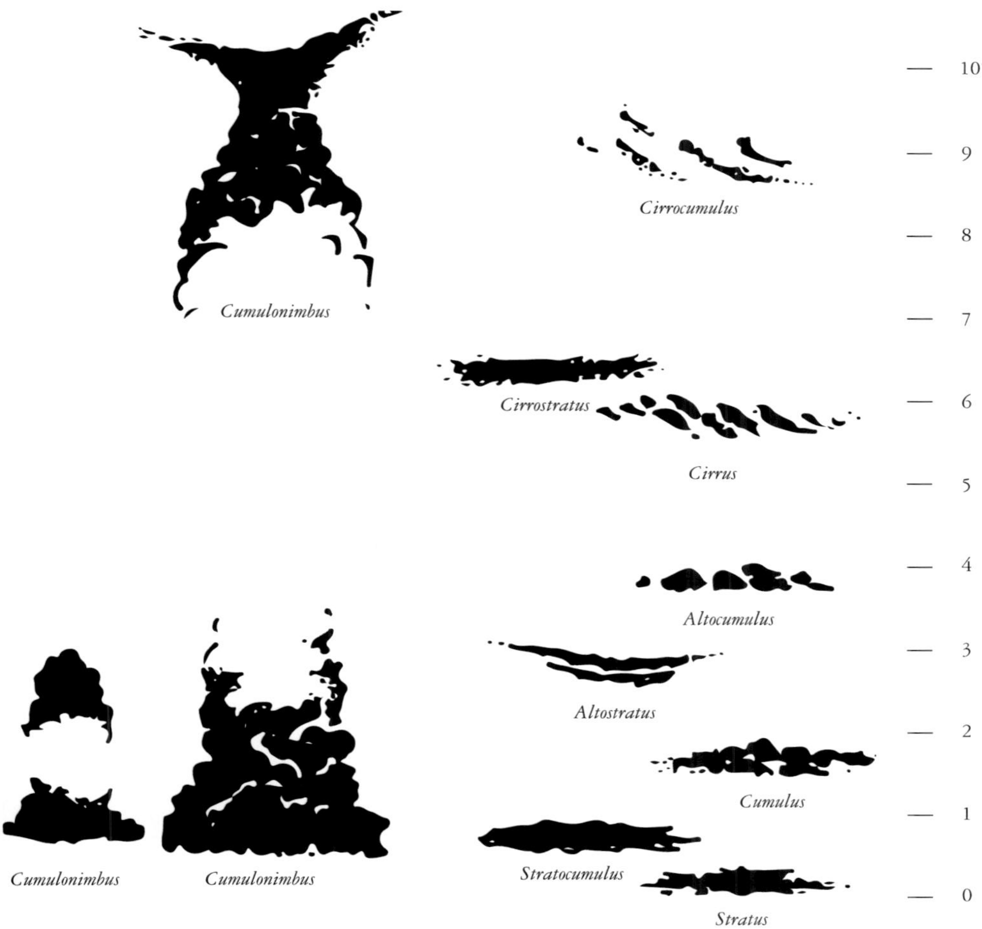
10
9
8
7
6
5
4
3
2
1
0
Cirrocumulus
Cumulonimbus
Cirrostratus
Cirrus
Altocumulus
Altostratus
Cumulus
Stratocumulus
Stratus
Cumulonimbus
Cumulonimbus

Andean hills welcome the coolness of the nocturnal clouds. Could they not be harnessed, forcing them to turn into the rain? Then, the engineers build a tightly meshed net to catch the water like fish. It is a fishing village. According to the principles of bionics, the plot is inspired by the structures of a canary tree. The tree 'rains,' while the dry region of the southern coasts where it lives never sees water. Its architecture and the imbrication of the leaves are such that they condense the humidity of the air. In Chugungo, the cloud gardener 'rakes' the condensed water on nets stretched up to seven meters high. Through curved gutters, the water is collected in large tanks. The village comes back to life, and the inhabitants return. You can even see flowers under the houses.

Every gardener—even those in regions where it rains—adapts to the clouds, supervises them, and doses his care according to the abundance or scarcity of meteors.

The first 'cloud' to come to our aid is the dew. It is not a formed cloud—with certain edges, as Lamarck would say—fat and limited to the upper regions of the troposphere. Although it is not a cloud at all, it works like one. Dew is produced as soon as the soil temperature drops to the point of starting the condensation of the water vapor contained in the layer of air with which it is in contact. To produce the same result, gardeners resort to a form of misting called 'sprinkler irrigation.' In Limousin, France, gardens are sprinkled with dew from mid-August. Barring exceptional heat, adding more water during the day is pointless. The absorption of water through the aerial tissues of the plant is just as necessary as that of the root tissues. The dew appears in the morning after a cool and clear night. Water droplets easily coalesce on ground obstacles as they bob in the air without settling. By raking the grass with a ladle, we can collect water like the gardeners of Atacama.

In general, plants 'get organized' with the climate. Massive watering, necessary for vases and other containers, has no reason to occur in the garden. The presence of species occurs in relation to the microclimatic sectors of the territory. The base of a tree is dry. Some species—periwinkles and cyclamen—resist this dryness. Wanting to put others that do not have the aptitude to tolerate this 'climate,' such as cyclamen or periwinkle—whatever their origin—forces artifice. Watering, watering, the programmer watering. Everything is underground, invisible, and effective. The tree, disturbed by this climate, rots and dies. Whether or not it falls on a house or the street during a storm depends on the butterfly effect of the moment. In an article entitled 'My dog was right,' Yves Delange stated how integrated watering, without criteria, has contributed to the disappearance of rare botanical species with heterogeneous needs. His dog did not like the hydraulic reel, an archaic and comical object that randomly distributes water.

While the pump irrigates, the gardener is occupied with other things. At this point, the question arises: Should the gardener care for something else?

However, he must learn how much to distribute in one case and how much to limit himself in another. He is increasingly studying water distribution according to the species' needs. As in so-called precision agriculture, these montages require a sophisticated and expensive material to do roughly what nature does for us for free. In summer, you have to let the lawn turn yellow.

Acknowledgments

This contribution consists of two parts. The abstract and introductory paragraphs 'Oceans of Clouds. Drops of Fog,' 'The Rain Dance,' and 'Learning from Nature' were written by Laura Cipriani. The paragraph 'Clouds' is the work of Gilles Clément and was translated by Laura Cipriani. The authors wish to thank Mr. Tonka of the Sens&Tonka publishing house for allowing the reproduction of the text from the book *Nuages*.

References

Authos (no date) *Neboulosus Mist Architecture.* Available at: <https://authos.ch/nebolosus-mist/> [Accessed 20 January 2023].

Azua-Bustos, A., Urrejola, C. and Vicuñ, R. (2012) 'Life at the dry edge: Microorganisms of the Atacama Desert.' *FEBS Letters,* 586, pp. 2939–2945.

Bhushan, B. (2020) 'Design of water harvesting towers and projections for water collection from fog and condensation.' *Philosophical Transactions A*, 378, pp. 1–37.

Chaves, M. (2010) 'Rain dances in the dry season: Overcoming the religious congruence fallacy.' *Journal for the Scientific Study of Religion*, 49 (1), pp. 1–14.

Clément, G. (1999) *Le Jardin Planétaire: Reconcilier l'Homme et la Nature Relié.* Paris, Albin Michel.

Clément, G. (2021) *Nuages.* Paris, Sens&Tonka.

Correggiari, M., Castelli, G., Bresci, E. and Salbitano, F. (2017) 'Fog collection and participatory approach for water management and local development: Practical reflections from case studies in the Atacama drylands.' In: Ouessar, M., Gabriels, D., Tsunekawa, A. and Evett, S. eds. *Water and Land Security in Drylands: Response to Climate Change.* Cham, Switzerland, Springer International Publishing, pp. 141–158.

Diller Scofidio+Renfro (no date) *Blur Building. Swiss Expo 2002, Yverdon-les-Bains, Switzerland.* Available at: <https://dsrny.com/project/blur-building> [Accessed 20 January 2023].

Ebner, M., Miranda, T. and Roth-Nebelsick, A. (2011) 'Efficient fog harvesting by Stipagrostis sabulicola (Namib dune bushman grass).' *Journal of Arid Environments*, 75 (6), pp. 524–531.

Eshel, G., Araus, V., Undurraga, S., Soto, D.C., Moraga, C., Montecinos, A., Moyano, T., Maldonado, J., Dıaz, F.P., Varalad, K., Nelson, C.W., Contreras-Lopez, O., Pal-Gabor, H., Kraiser, T., Carrasco-Puga, G., Nilo-Poyanco, R., Zegara, C.M., Orellana, A., Montecino, M., Maass, A., Allende, M.L., DeSalle, R., Stevensonk, D.W., Gonzalez, M., Latorre, C., Coruzzi, G.M. and Gutierrez, R.A. (2021) 'Plant ecological genomics at the limits of life in the Atacama Desert.' *Proceedings of the National Academy of Sciences*, 118 (46), pp. 1–11.

Gurera, D. and Bhushan, B. (2020) 'Passive water harvesting by desert plants and animals: lessons from nature.' *Philosophical Transactions A*, 378, pp. 1–23.

Hamilton, W. and Seely, M. (1976) 'Fog basking by the Namib Desert beetle, Onymacris unguicularis.' *Nature*, 262, pp. 284–285.

Henschel, J.R. and Seely, M.K. (2008) 'Ecophysiology of atmospheric moisture in the Namib Desert.' *Atmospheric Research*, 87 (3–4), pp. 362–368.

Hunchuck, E.M., Ferrari, M. and Cheng, J.C. (2021) 'Prologue to the sky river.' *The Avery Review*, 53, pp. 209–248.

Ju, J., Bai, H., Zheng, Y., Zhao, T., Fang, R. and Jiang, L. (2012) 'A multi-structural and multi-functional integrated fog collection system in cactus.' *Nature Communications*, 3, pp. 1–6.

Laaksonen, A. and Malila, J. (2022) 'Ice nucleation.' In: Laaksonen, A. and Malila, J. eds. *Nucleation of Water*. Amsterdam, Elsevier, pp. 209–248.

Lamarck, J.B. (1805) 'Nouvelle définition des termes que j'emploie pour exprimer certaines formes de nuages qu'il importe de distinguer dans l'annotation de l'état du ciel.' *Annuaire Météorologique Pour l'An XIII de la République Française*, 6, pp. 112–133.

Latz+Partners (no date) *CHAU. Fog Garden, Chaumont-sur-Loire, FR*. Available at: <https://www.latzundpartner.de/en/projekte/besondere-orte/nebelgarten-chaumont-sur-loire-fr/> [Accessed 20 January 2023].

Merton, R.K. (1968) *Social Theory and Social Structure*. New York, The Free Press.

Mitchell, D., Henschel, J.R., Hetem, R.S., Wassenaar, T.D., Strauss, W.M., Hanrahan, S. A. and Seely, M.K. (2020) 'Fog and fauna of the Namib Desert: past and future.' *Ecosphere*, 11 (1), pp. 1–40.

Mooney, H.A., Gulmon, S.L. and Weisser, P.J. (1977) 'Environmental Adaptations of the Atacaman Desert Cactus Copiapoa haseltoniana.' *Flora*, 166 (2), pp. 117–124.

Morales, A.G., Ojeda, A.A.R. and Torres, S.H. (2015) 'El cultivo del viñedo como recurso turístico cultural: El caso de La Geria (Lanzarote. Islas Canarias, España).' *Papeles de Geografía*, 61, pp. 109–121.

Roth-Nebelsick, A., Ebner, M., Miranda, T., Gottschalk, V., Voigt, D., Gorb, S., Stegmaier, T., Sarsour, J., Linke, M. and Konrad, W. (2012) 'Leaf surface structures enable theendemic Namib desert grass Stipagrostis sabulicolato irrigateitself with fog water.' *Journal of the Royal Society Interface*, 9, pp. 1965–1974.

Schaefer, V.I. (1946) 'The production of ice crystals in a cloud of supercooled water droplets.' *Science*, 104, pp. 456–459.

Theurer, S.J., Lissoni, A. and Duguet, A. eds. (2023) *Fujiko Nakaya*. Berlin, Hatje Cantz.

Vonnegut, B. (1947) 'The nucleation of ice formation by silver iodide.' *Journal of Applied Physics*, 18, pp. 593–595.

Wang, G., Deyu Zhong, D., Lia, T., Zhang, Y., Meng, C., Zhang, M., Song, X., Wei, J. and Huang, Y. (2018) 'Study on sky rivers: Concept, theory, and implications.' *Journal of Hydro-environment Research*, 21, pp. 109–117.

13 | Cultivating Water

From the Dew to the Sky Vault

Catherine Mosbach

Design Project Data

Location
Taichung | Taiwan | Asia
Project Team
Mosbach Paysagistes | Philippe Rahm Architectes | Ricky Liu and Associates, Architects+Planners
Date
2011 | 2021
Coordinates
24° 9' 0" N | 120° 40' 1" E
Area
0.674 km^2
Length
2.4 km
Investment
80.3 million euro

The inhabitants of Taichung, Taiwan, were deprived of drinking water for several months in 2021 due to a bloodless water table, deforested mountain slopes, and channeled streams, with all the water flowing directly into the ocean. This chapter presents Phase Shifts Park, the keystone of a 250-ha urban development northwest of Taichung's historic center, which offers a new paradigm for city dwellers. New water is cultivated by the biomass introduced with the project. The water cycle relays life from the depth of the roots to the foliage, which evaporates the surplus in the form of oxygen into the sky. The amplitudes initiated by this park are rare in the sense that they introduce 93% of porous soil into a new centrality, representing 80% of the possible right-of-way available at high monsoonal waters. Large and small holes and filaments invite microorganisms to the mother rock, releasing the mineral salts essential to life. Water is no longer an element that must be contained or protected against. This chapter shows that water is the source of life, inviting inhabitants to pay close attention to this resource, which is the keystone of a climatic *balancing act* to benefit the continuous regeneration of life on earth.

'Cultivating' water in Taiwan

Water explores a multi-dimensional constellation. It questions scales and temporalities from the most infra-mycological to the most macro-planetary. It is a matter of landscape in the sense that it is part of the common good, and its treatment in our design process is a small stone in the edifice of required attention. Although some briefs are more conducive than others, all should 'cultivate' water. It is an optimal linking of the transmission belts from deep in the ground to high in the atmosphere. At two sites in Asia—0.7 km^2 in Taichung, Taiwan, and 500 m^2 in Ulsan, South Korea—the relationship to water directs the design as fundamentally as gravity binds us to the ground.

Taiwan, the east coast of which is exposed to the Kuroshio Current—one of the most significant marine currents in the world—evolves under a humid tropical sky. Its soil relays the seismic movements of its volcanic resurgences. These two parameters condition the relationship of Taiwan's inhabitants to their territories. On the one hand, they take refuge in air-conditioned interiors and, on the other, in places sheltered from volcanic tremors. Their relationship to the environment reflects a distancing, a reserve. A former airport north of the historic center offers the opportunity to create new links between living elements and populations. In Phase Shifts Park, water is this active link that articulates all those dimensions with the possibility of a sensitive link co-produced between elements and humans. The design of this park is a catalyst for these assets. It can only achieve something when deeply untying the paths of symbiosis between self-regulating dynamic systems.

Here, the water infrastructure produces life from a vast folding of the soil from north to south. Micro- and macrotopographies initiate water capture and the emergence of living environments. Lineaments, holes, and extents retain water that, when in contact with the air, welcomes the microbial communities of plants and soils—a symbiotic relay that produces amino acids and oligo-elements, the pillars of life. Rainwater irrigates an 'epidermis' soil that retains it where it falls, down to the water table. The volume of roots, rootlets, and mycorrhizae leads it to the bedrock, which in turn releases the mineral salts that sustain life in all its forms. Soil textures and canopy volumes precisely calibrate a runoff coefficient according to the particularities of each territory. The active tissues of phytoremediation plants filter the water from urban facilities upstream of their liquid-vapor transformation cycle to the ocean clouds. The leaf-root biomass of the plant strata preserves and generates new water, which is relayed by the phenomenon of the water lift of the mycorrhizal hyperfluidity network. This network distributes the useful water reserve to the plants and soils by condensation-evaporation up to the atmosphere, thus recharged with oxygen by the product of photosynthesis.[1]

This stratification in depth and extent facilitates the versatility of recreational appropriations. Water morphing allows for the coverage of approximately 80% of submerged areas during the monsoon season. Water thus choreographs the assets of human activities without distancing and defiance. It draws the morphological potentials between wet seasons of saturation and dry seasons when the root volume returns the element of water to the surface by capillary action to compensate for the arid sequences. A pathway out of water guides pedestrians, whatever the hazards of the water seasons.

From the largest to the smallest, Lost in Transition Park is reminiscent of a steep relief that hosts the bed of the Taehwa River in its heart—a low point of runoff from the surrounding mountainsides where liquids and solids intertwine, each carrying their biological assets. Before its containment by human settlements, the Taehwa River exhibits its original course to the mouth of Tsushima Strait. The plasticity of its meanders reflects the obstacles it crosses, showing the riverbed of evidence that it fights through the ages and seasons. The deposits it carries are rubbed along its passage, transforming a diversity of soil environments. Between the highest areas of South Korea—the peaks of Taebaek with their reservoirs of water, seeds, and fresh air—and the lowest—the seabed with its reservoirs of continuous cultures and migrations—this fragment of the garden is developed.

The dynamics of the flows and the power of morphogens are pregnant, from the atmosphere with its whirlwinds of air and differential attractions between elements to the seabed with masses of water, fish, and fishponds in transit, into the infra-thin bodies with solid-liquid defectors of our intimacies. Since the 1970s, researchers (Schwenk, 1978; Cloos, 2015) have revealed that the fluid dynamics of mountainous reliefs (such as the Alps)—like those of human organs (such as the heart)—activate similar flows and life processes between solids and liquids.

Here, the water sneaks into the folds and crevices on the riverbank. Foliage filters stripes of light to the ground reflections. Mosses cling to the asperities of the deposits. Sheers protect an interlude of relaxation amidst the plants. Shoots of the year and their umbels communicate the vigor of fertile soils. The tracings of the men who build up the garden invoke the tradition of petroglyphs flush with the rocks of Ulsan, which exposes terrestrial food in gratitude for their benefits. Lost in Transition retains the alliances between *humanities* and *environments*, relaying beyond the strictly human temporalities. This small garden offers the possibility of representing these phenomena in a riverbed that is now constrained but which relays these dynamics to perpetual formations-reformations and initiates *appearances* that capture decaying and fading memories of these essential to the regeneration of the living.

The landscape architect initiates places by putting the necessary assets engaging the fabric of life into optimal interactions. This hybridization sensitizes populations to evolve in contact with non-human alterities, not against them or by induced subordination.

Figure 13.1 | Phase Shifts Park, Taichung, Taiwan. Photograph by Victor Choao Wu.

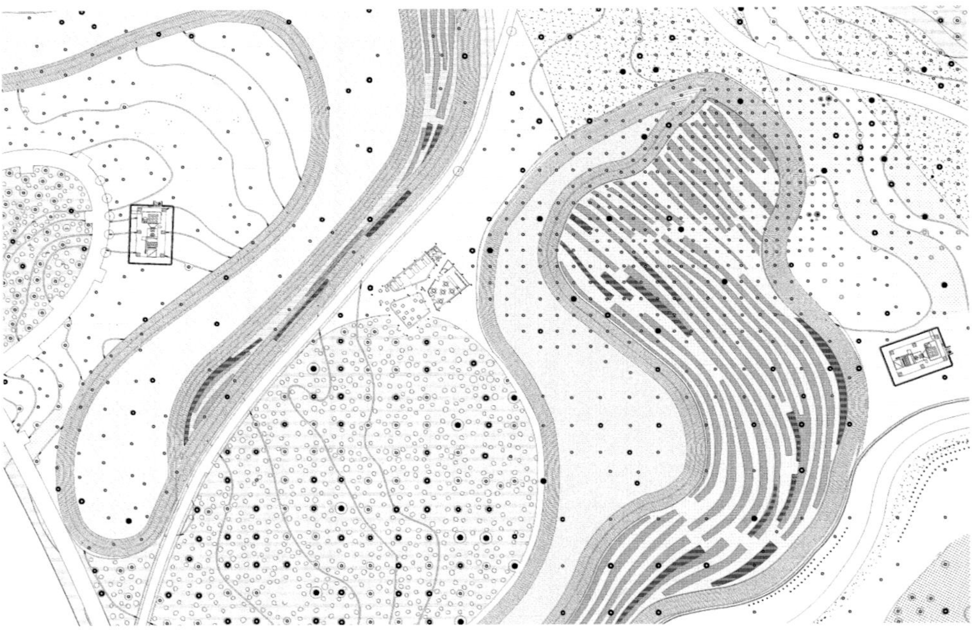

Figure 13.2 | Phase Shifts Park, Taichung, Taiwan. Image by Mosbach Paysagistes and Philippe Rahm Architectes.

Figure 13.3 | Phase Shifts Park, Taichung, Taiwan.Photograph by Victor Choao Wu.

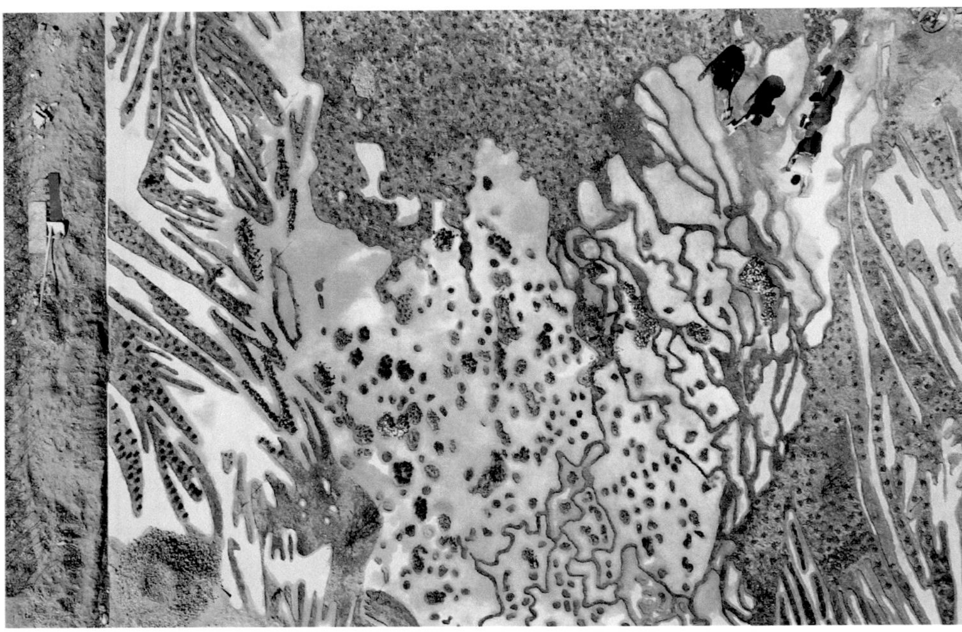

Figure 13.4 | Lost in Transition Park, Ulsan, South Korea. Photograph by the Korean Institute of Garden Design.

Note

1 | The word *photosynthesis* comes from the Greek *phos* (light), *syn* (together), and *tithenai* (to combine). The process by which plants synthesize organic matter with light energy by absorbing carbon dioxide (CO_2) from the air and releasing oxygen (O_2). Plants and bacteria can make energy and produce their constituent parts from a few particles of sunlight. They are said to be autotrophs (i.e., they can make their substance from inorganic material). During photosynthesis, a leaf emits not only electrons but also protons, which partly recombine with oxygen and create water. The plant produces water that does not come from precipitation; water that did not exist before.

References

Cloos, W. (2015) *The Living Origin of Rocks and Minerals*. Edinburgh, Floris Books.

Schwenk, T. (1978) *Sensitive Chaos*. New York, Schocken Books.

14 | Cloudscapes

Transcending Climatic Boundaries

Nadir Abdessemed and Christine von Raven

Design Project Data

Location
Venice | Italy | Europe
Project Team
Transsolar KlimaEngineering | Tetsuo Kondo Architects
Date
2010
Coordinates
52° 00' 24.01" N | 4° 21' 20.02" E
Area
900 m²
Length
40 m
Investment
n.a.

Cloudscapes was a microclimatic installation featuring an accessible indoor cloud at the 2010 Architectural Biennale in Venice. It exhibited continuously changing temperature layers. Since the cloud redefined space through climate, this represents opportunities for understanding architecture and the way buildings engage with the environment thermodynamically, with the hope of spaces being balanced by natural forces and a more harmonic connection between humans and climate both indoors and outdoors.

Transcending Climatic Boundaries

Hope adopts an optimistic attitude in an otherwise unpromising context. It may yield positive momentum: A self-fulfilling prophecy that creates a path to a better future. In architecture, hope transcends design limitations. It blurs constraining boundaries and makes a big challenge appear lighter, sometimes turning the impossible into something feasible.

Cloudscapes is an installation meant to bring a real cloud into a historic building at the Venice Biennale, the Corderie del Arsenale. It is based on optimism and hope since the artificially created atmosphere enables the ostensibly impossible. It represents the immateriality of spatial boundaries and defines a weightless yet atmospherically and thermally distinct space hovering above people.

Clouds are important elements of our planet's atmosphere, framing outdoor space and filtering the sunlight. They are the visible part of the terrestrial water cycle, carrying water—the source of life—from ocean to land. Clouds naturally sustain themselves in equilibrium, embodying and releasing solar energy. Although the experiment is based on an artificial climate, Cloudscapes is far from dreaming about terraforming, where humans control the weather on a global scale. It is rather a *physical model* of a cloud, focusing on the interaction of warm, humid layers of air with colder ones (Abdessemed, 2018; Abdessemed and Schuler, 2010). By allowing vapor to assume the form of droplets in the right atmosphere, Cloudscapes works with water in different aggregate states. This allows visitors to experience the dynamics of the cloud landscape itself; for example, children running in circles can cause a descending cloud vortex that stabilizes itself after some time. The interaction between climate and people forms a distinct topography as the installation invites visitors to walk through a temperature-humidity *continuum*.

Notably, 419 ppm is an abstract number describing a part of the atmosphere that we cannot see but do breathe. Cloudscape's meaning for architecture lies beyond a romantic description of an ephemeral landscape. Despite being an artistic representation, it has physical meaning for the nature of architectural spaces at all scales. Climate and its underlying thermodynamics connect all scales of life, be they designed by nature or humans, ranging from our global atmosphere to regional climate zones and urban microclimates, to landscapes, buildings, rooms, and even the human metabolism. Ironically, the root of our climate crisis lies in our creation of envelopes of steel, concrete, and glass to protect us from the weather. Aiming for more comfort within artificially created microclimates, we consume energy and emit CO2 into the atmosphere. It is exactly this mechanism that alters the climate on a global scale, giving rise to our climate crisis as buildings contribute nearly 40% of global greenhouse gas emissions (Global Alliance for Buildings and Construction, International Energy Agency and the United Nations Environment Program, 2019).

Cloudscapes dissolve the concept of isolated climatic conditions that a building enforces. While a building conventionally forms one *discrete* climate separated by a substantial and harsh boundary—the building envelope or a wall—architecture can embrace a more continuous path of different climates. Cloudscapes exemplify these *continuously* changing conditions. It invites us, as humans, to abandon our tight thermal comfort expectations by creating spaces of different temperatures and humidity merging into each other while still displaying well-defined distinct states, where water is either in an invisible gaseous form or in a droplet phase. Experiments involving climatic buffer spaces, sometimes referred to as 'third spaces' or 'mid-doors' (neither indoors nor outdoors), also exist in permanent projects. Zaryadye Park in Moscow and the terrace of the Academy Museum of Motion Pictures in Los Angeles create thermal buffer spaces that are 'open' (can be accessed without doors) and exhibit climates that are somewhat protected from the outdoors, yet not artificially conditioned. Other approaches range from a simple winter garden—such as the glass envelope covering the historic Strasbourg train station—to open landscapes that continuously change, shaping the local microclimate and human thermal experience (e.g., Benedito, 2022). By showing a continuous connection between dynamic climates under more flexible conditions, Cloudscapes creates hope for an antidote to humans controlling climate with fossil fuels, releasing energy that has been harnessed millions of years ago.

It must be stated that it requires a tremendous amount of technology, not to mention energy, to create this special atmosphere saturated with water. Yet, it is this very revelation that reminds us of the power of atmospheric dynamics: The sun, the water, and the air create very distinct different climates in space and time. It is the task of architecture to harness these forces, and these examples give us hope that we will collectively become increasingly conscious of the energetical and microclimatic interdependence between the global and human scales.

Figure 14.1 | Cloudscapes by Transsolar and Tetsuo Kondo Architects. Architectural Biennale Venice 2010. Visitors experience the atmosphere defining the cloud layer via a continuous ramp and find themselves below, in between, and above the cloud. Photograph by Tetsuo Kondo.

Figure 14.2 | Below the cloud, the air is dry and therefore 'heavier' than the humid atmosphere hosting the cloud. Photograph by Tetsuo Kondo.

Figure 14.3 | Above the cloud, the air is warmer and lighter than the fully saturated cloud layer. Photograph by Tetsuo Kondo.

References

Abdessemed, N. (2018) 'On the nature of thermodynamic models in architecture and climate.' *Humanizing Digital Reality*. Singapore, Springer Nature, pp. 85–91.

Abdessemed, N. and Schuler, M. (2010) *Cloudscapes*. Stuttgart, fmo publishers.

Benedito, S. (2022) *Atmosphere Anatomies: On Design, Weather, and Sensation*. Zürich, Lars Müller Publishers.

Global Alliance for Buildings and Construction, International Energy Agency and the United Nations Environment Program (2019) *2019 Global Status Report for Buildings and Construction: Towards a Zero-Emission, Efficient and Resilient Buildings and Construction Sector*. Available at: <https://iea.blob.core.windows.net/assets/3da9daf9-ef75-4a37-b3da-a09224e299dc/2019_Global_Status_Report_for_Buildings_and_Construction.pdf> [Accessed 19 July 2023].

Conclusions

15 | Choreography in Landscape Practice

Agency, Intention, and the Beauty of a World in Motion

Emily Knox and Rob Holmes

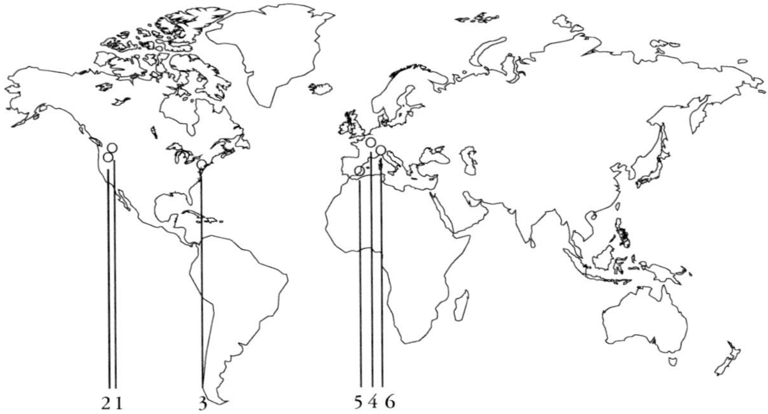

Design Project Data

1 | Freeway Park
Location
Seattle | Washington | USA | North America
Coordinates
47° 36' 36.184" N | 122° 19' 49.502" O

2 | Open Space Sequence
Location
Portland | Oregon | USA | North America
Coordinates
45° 30' 45.816" N | 122° 40' 44.902" O

3 | Manhattan Square Park
Location
Rochester | New York | USA | North America
Coordinates
43° 9' 13.812" N | 77° 36' 7.289" O

4 | Parc du Sausset
Location
Villepinte | France | Europe
Coordinates
48° 57' 42.877" N | 2° 30' 41.088" E

5 | Valencia Central Park
Location
Valencia | Spain | Europe
Coordinates
39° 27' 27.332" N | 0° 22' 42.753" O

6 | Mühlenbachstrasse Courtyard Wall
Location
Zurich | Switzerland | Europe
Coordinates
47 ° 21' 57.347" N | 8° 32' 57.164" E

Making landscapes is not only for landscape architects. Many others, such as dredge operators, ranchers, foresters, and restoration ecologists, are also engaged in this work. These fields have their own practices, which are often quite different from those of landscape architects. Some of these other practices, which involve collaboration between people, animals, plants, and landscape processes, can be understood as *choreography*. Choreographers directly intervene in and through landscape processes such as sedimentation and seed dispersal, repeatedly interact with the landscapes they are working in, make decisions in reaction to their observations, and use tools such as remote sensing and embodied perception to pay careful attention to how landscapes evolve.

Choreography has a long history as an artistic practice, and landscape architects have occasionally adopted the term. This chapter proposes that choreography has the potential to invigorate landscape architectural practice for three reasons. First, because it is fundamentally collaborative, balancing intention with the agency of all participants in a manner that is directly aligned with the concept of (co)design. Second, because it is suited to making landscapes at much larger scales than traditional landscape construction, which climate change and other contemporary environmental crises will require. And third, because choreography is simultaneously technical and poetic, deploying the generative potential of ecological processes to imagine and construct landscapes.

Landscape Makers

The rancher rides with the herd, directing cattle across rangeland; the cattle graze, and plant communities evolve in response. The next year, the rancher will consider where the best grazing lies, and move the herd accordingly. The dredging crew returns to the same navigation channel year after year; they dredge out sediment and place it in shallow water, where it piles up to form islands. Marsh grasses grow on the new land. Cattle, marsh grasses, dredge operators, ranch hands, rangeland forbs, and dredged sand are all engaged in the work of making landscapes: Sometimes purposefully, sometimes unintentionally—but making landscapes either way.

When this type of collaborative work between people, animals, plants, soil, water, and other landscape agents is organized with intention, that organization can be understood as *choreography* (Galí-Izard, 2005; Lokman, 2017; Cantrell, Zhang and Liu, 2021; Girot, 2021; Knox, 2022; Holmes et al, forthcoming). In a landscape context, choreography is the intentional orchestration of patterns of movement and change to shift the trajectory of a landscape in a desired direction. Landscape architects are increasingly interested in design practices that, similar to choreography, work with time, process, and landscape change (Dee, 2010; Milligan, 2015; Raxworthy, 2019; Geffel, 2021). As works like this volume attest, landscape architects are also concerned with participatory and collaborative (co)design methods (Hester Jr., 2010; Hirsch, 2014; De la Peña et al, 2017; Solan, 2017).[1]

Meanwhile, although the landscape makers we have named, such as dredge operators and ranch hands, would not call their work choreography—in fact, their work is more often described in the rationalized terms of scientific management (Rosati et al, 2004; Briske, 2017)—choreographic qualities are latent in their work. Choreographers make landscapes, but they make landscapes in very different ways from traditional landscape architecture practitioners. By borrowing and adapting the choreographic methods of these landscape makers, landscape architectural practice can be reinvigorated in ways that are particularly well-suited to contemporary conditions, including the climate crisis.[2]

Controlled Landscaping

The world of landscape architecture is not like the worlds of rangelands and sediment.[3] Rather, the traditional methods and means of landscape architecture—landscape design, construction, and maintenance, as typically practiced—are oriented toward control. Herbicides eliminate unwanted volunteers, while mowers and weed-whackers trim grass and groundcovers to assigned heights. Nursery inventories ensure that the plants placed are the plants selected, down to caliper, pot size, and bloom color. Rebar reinforcement and footing widths are engineered to prevent

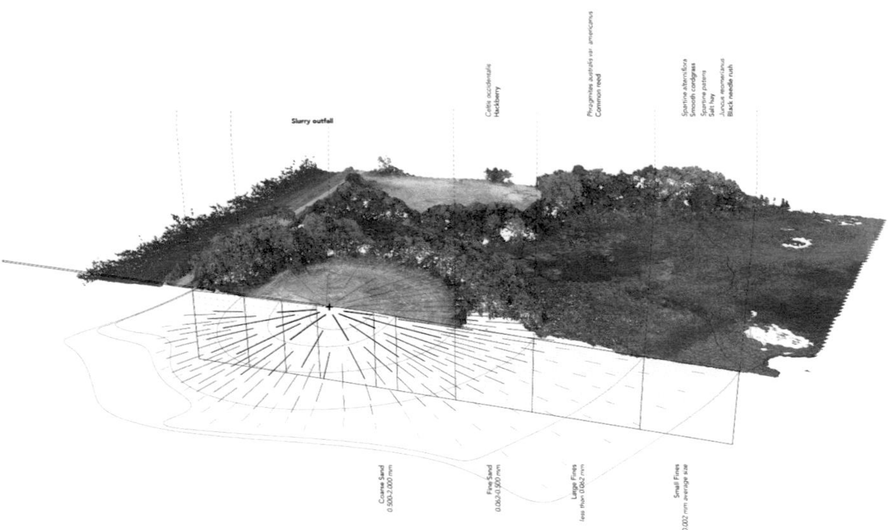

Figure 15.1 | A slurry of dredged material has been piped onto an island on South Carolina's coast, where it forms a mound that is self-sorted by grain size and, as vegetation responds to the resulting differences in elevation, wetness, and soil, rings of distinct plant communities form. Drawing by Rob Holmes and Valerie Friedmann.

cracking, deformation, and slumping. Across larger territories, excavators carve channels to shape the movement of water, while erosion control mats and hardened surfaces are used to resist it. Moreover, technologies such as infiltration planters and retention basins work to keep water fluctuations within determined parameters.

One effect of this is that landscapes are flattened and diminished, rendered in monotones rather than rich diversity. Correspondingly, landscape architecture is limited to a weakened palette. Instead of meadows that harbor a multitude of insect species, we plant lawns of zoysia and crabgrass dotted with ornamental shrubs. Instead of forests with dozens of hardwood tree species forming an ever-shifting matrix of patches and stands, we compose groves, allees, and bosques (not to diminish the wonder of a bosque!) of the same handful of tree species available from commercial nurseries. Stone cut by saws and water jets, and thousands of meters of concrete with almost precisely the same aggregate mix, instead of the fluid irregularities of karst limestone or the subtle gradients of cobble, gravel, and sand sorted by the action of a river. Gradients between wet and dry become irrigation schedules and 'water features.' Notably, planting design informed by ecological understanding and the aesthetics of growth, like that of Nigel Dunnett, Teresa Galí-Izard, or Piet Oudolf, and an inventive approach to the materials of landscape construction, as in the work of Vogt Landscape Architects or Teresa Moller's studio, can mitigate some of this. However, working only with the tools of traditional landscape construction is aesthetically limiting.

Another effect is a weakened capacity for the landscape to regenerate, self-organize, and support life. Planting plans and construction documents not only specify trees and details but also a compressed set of potential relationships and processes. Much of landscape architects' work, and the work of those charged with its maintenance, is designed to stifle and suppress emergent landscape properties. When new plant species are hastily removed to maintain the originally specified character of the work, or when mulch and herbicide are applied, successional processes are slowed or stopped. When water is hurriedly drained from sites, it is no longer able to pool or seep, where it might generate new conditions. The resulting landscapes have a limited set of trajectories, cut off from the resources and energy that would typically bring about change. They are unable to conjure novelty. Work like this is governed by human limitations—the limitations of both our imaginations and our physical capacity to create.

Underlying these effects and the tools that (re)produce them is an ideology of control centered on modernist concepts such as scientific management, rationalization, and efficiency (Bélanger, 2016; Shallat, 2010; Sayre, 2017; Hays, 1999).[4] In ecological science, these concepts are associated with an outmoded paradigm that understands nature and natural systems as fundamentally balanced, linear, and progressive (Cook, 2000; Wu and

Loucks, 1995). This outmoded paradigm has been radically complicated by an appreciation for the utility of disorder and the dynamic qualities of ecological phenomena. In the sciences more broadly, there has been an associated shift toward 'post-normal science (...) based on unpredictability, incomplete control, and a plurality of legitimate perspectives' (Funtowicz and Ravetz, 1994, p. 1881). Modernism's certitude and control have proven both inadequate for the realities described by these newer paradigms and frequently counter-productive, introducing cascades of unintended consequences into systems that undermine their designers' goals (Holmes, 2020).

By contrast, choreography is fundamentally collaborative: It balances direction and intention with openness and the agency of all participants. It works in gradients (an action followed by a response by other agents; observation and another action; further response...) rather than binaries (either constructed or not constructed). When we allow other agents into the design process, choreographing roles and avenues of expression for them, we open ourselves to the possibility of being surprised by what we make together. We can escape the modernist desire to transcend our human limitations by increasing our power and control, instead acknowledging our limitations while working within our finitude, our necessarily incomplete knowledge, and our lack of control to (co)design with other agents and achieve what none of us could alone.

Scale, Tools, and Transformation

Landscape architecture today is not practiced, though, in a stable environment, where we might have the luxury of gradual and gentle experimentation. We now practice amid a global destabilization of landscape processes. The accelerated, transcontinental migration of species via human trade routes has produced novel ecosystems, even on remote islands. The cycles of primary nutrients such as phosphorous and nitrogen are now largely industrialized. Moreover, carbon emissions are transforming the planet's atmosphere, warming its climates beyond historic precedent, and igniting interconnected crises for both humanity and the other species of Earth.

Observing the rapid temporal pace and vast spatial scale of human impacts some two decades ago, French landscape architects Christine Dalnoky and Michel Desvigne noted the following in an interview:

> 'The European landscape has changed more in the past thirty years than in all the preceding period... Of course, it is legitimate for European landscape architects faced with the scale of change in the landscape to dream of being involved in the processes. It is equally justifiable to claim an important social role for landscape architecture on the basis of these changes. But paradoxically this is not the case, this role does not exist.'
>
> (Schröder, 2001, p. 85)

Figure 15.2 | Cows at Alderspring Ranch herded across a 200-km^2 allotment in the Salmon-Challis National Forest, Idaho, USA. The disturbance they produce induces landscape change at a large scale. Photograph by Melanie Elzinga.

The interviewer, Thies Schröder, went on to note that 'Desvigne and Dalnoky have put the marginal role of landscape architecture into figures: Every French landscape architect works on an area per year corresponding to an eightieth of the land changed by forest fires, or a thousandth of the area that becomes urbanized. Working on these figures, it would take 2700 years merely to redesign the landscape of all the agricultural land that has currently been set aside [in France].' The landscape architects draw out the obvious implication: 'We neither define nor control the future of the landscape' (Schröder, 2001, pp. 85–86).

Their critique is important, and it applies to every continent.[5] Climate change and other contemporary environmental crises demand large-scale responses, and landscape architecture, as traditionally practiced, is ill-suited to providing those responses. There is a fundamental mismatch between the scale of landscape issues and the capacity of the tools that we use. Artifacts such as construction documents and planting plans are logistically and financially futile in the context of a rangeland or river delta measuring tens or even hundreds of thousands of hectares.

These static tools beget static responses, passive objects in the landscape. If they are ever to meaningfully participate in addressing the changing climate, water crises, or energy transitions, the language and tools of the discipline will need to evolve and expand. In the process of incorporating these new tools, new methods must also be carefully crafted and sharpened to avoid losing the existing richness of landscape architecture, which (at best) incorporates a much fuller range of values than the managers, scientists, and planners who currently make decisions about large landscapes. Over the past 30 years, we have tinkered with our disciplinary values, generating new modes of thinking about and representing large, complex landscapes such as rangelands and river deltas (Corner, 1997; Mathur and da Cunha, 2009; Coomans, Cattoor and De Jonge, 2019; Milligan, 2022). However, we have made relatively little progress in exploring methods for taking action.

In recommending choreography as a mode of (co)design, we are asserting that we have much to learn from landscape practitioners such as rangeland managers and sediment choreographers, who are already working in ways that are suited to making (and re-making) landscapes at much larger scales than traditional landscape construction is capable of.

How do choreographers work? What can landscape architects learn from them?

Intervention and Processes

Choreographers intervene in and through landscape processes such as sedimentation and erosion, like root growth and seed dispersal.

In traditional landscape practice, the designer draws, specifies, and confirms that contractors have produced a landscape precisely according to a set of construction documents. An end result is identified, with designers and builders working as a team toward that end result. Bulldozers, excavators, and shovels move earth until it aligns with the contours of a grading plan. Plants are specified by species—often by precise cultivar, meaning that the plants in a designed landscape are often genetic clones of one another—and placed in locations that have been denoted on a planting plan. This mode of construction is limited in both its potential to produce diversity and its capacity for scale.

In contrast, the choreographer seeks to instigate change by meddling with the active processes that shape landscapes before and after human intervention. For instance, from planting to harvest, a forester returns to the same forest to intervene in processes of growth and succession. Trees are planted in tight rows to encourage competition and vertical growth. Depending on the species and location, foresters may return at some interval to deploy a disturbing agent like fire or herbicide, slowing succession and suppressing species that compete with emerging saplings. At a larger timescale, they return to conduct thinning operations, adjusting competition between trees by manipulating light and space availability. By engaging the processes of the forest, they work to shift forests' trajectories toward the most efficient growth of the species under cultivation.

Similarly, a rancher deploys their herd to graze and consequently produces disturbance. That disturbance begets a corresponding response from the landscape, and other processes are set into motion. For instance, new patterns of water infiltration emerge in response to the imprints of hooves, and the novel pattern of water availability causes some plants to thrive while others decay. Over time, such processes aggregate to produce intricate spatial patterns across great distances, with the initial disturbance shaping (but not mandating) the particulars of why, when, and how those changes unfold.

Interaction and Engagement

Choreographers repeatedly interact with the landscapes they are working in. They engage landscapes and the materials, beings, and processes that constitute them in continuous conversation, making decisions in reaction to the conditions they observe.

In traditional design practice, the work of the landscape architect ends once the drawing or construction process is complete. Ideally, the work of the choreographer might never end. Choreographers are fundamentally engaged with time. They understand that landscapes never finish their making and unmaking. Thus, choreographers also return again and again in continuous conversation with the materials, beings, and processes that constitute the landscapes in which they work. The frequency of interaction directly responds to the cycles of the landscape in question. A cold desert rancher returns summer after summer to the range, choreographing their herd in response to what they see before them as each new growing season unfolds. Dredging is scheduled in response to the particularities of depths, uses, and rates of sedimentation, which vary greatly. For example, a small channel used by recreational boaters may be dredged every few decades, while a deep commercial channel cut into a shallow estuary that receives water and sediment from a broad watershed may be dredged annually. The forester repeatedly returns and chooses when to burn, apply herbicide, thin, or harvest based on the state of the trees and competing vegetation.

This ongoing conversation with the landscape allows a type of responsiveness and open-endedness not possible in traditional models of design. Climate change promises that environmental flux will increase while becoming both less predictable and more extreme. While landscape practice might attempt to hold that flux at bay, the choreographer is already accustomed to responding to changes over and over again. For instance, in a dry summer on the range, a rangeland choreographer might respond by reconsidering the trajectories of the herd, shifting their speed, spread, and intensity to produce a larger but lighter impact. While scenarios can be anticipated and such responses can thus be planned ahead of time to some extent, implementing and adjusting them requires deep, continual engagement.

Observation and Embodiment

Since their observations shape their decisions, choreographers train themselves to pay close and careful attention to how landscapes evolve by using a variety of means ranging from remote sensing to directly-embodied perception.

In an estuary, for instance, suspended sediment concentrations may be high on one side of an estuary, with the marshes in that region rapidly accreting in response, thereby gaining elevation—perhaps even faster than the sea level is rising. In another portion of the same watershed, upland modifications to inflowing tributaries may block sediment from ever reaching the estuary's waters, causing sediment-starved marshes to rapidly erode. In addition to fieldwork, sediment managers can deploy a robust suite of remote sensing tools such as Laser Imaging Detection and Ranging (LIDAR), sonar, tracers, and sediment gauges, which allow them to observe and document emergent processes, patterns, and flows. Choreographers also work closely with scientists and modelers to understand the trajectories of landscape change at large spatial and temporal scales.

Yet the observations underlying choreography do not always require expensive, complex technologies. A rangeland choreographer requires a much different and more direct form of embodiment. Rangeland managers spend much of their time deep in the field, on horseback or All-Terrain Vehicles (ATVs), observing the vast landscapes they manage. Decision-making is based primarily upon this scale of deeply embedded, day-to-day embodiment. While technology is slowly becoming more accessible, decisions on the range are primarily memory-based, with ranchers often responsible for tens of thousands of acres. To choreograph safe routes for their herd across such vast territories, they must be able to read and recall plant compositions and densities, topographies, soils, and water sources. Year to year, they must also calibrate those readings to shifting climatic and weather conditions, which may alter those assemblages. Observations and decisions are made in real time.

Whether more technologically mediated or less so, such methods for observation and attention—if adopted by landscape architects—would force a corresponding questioning of disciplinary conventions. Designing with processes fundamentally challenges common methods, rebuking artifacts such as planting plans or construction documents, which become ineffective when working with uncertainties. Such drawings treat form as static; they imply that a single snapshot in time is adequate. Representational tools and techniques grounded in form as trajectory are needed. Time must become a central figure (or field) in our drawings. However, we must not think that merely representing the landscape as dynamic or acknowledging time in our drawings is sufficient. In choreography, drawings are always a tool for engaging the real phenomena of flux and transformation, and what is choreographed is ultimately actual agents acting in the landscape, not lines and symbols on a page or screen (Raxworthy, 2019).

Unruliness and Releasing Control

Choreographers are aware that the choreographed can be unruly: Sheep are directed to a field, but they make their own decisions about where within that field they graze; sand is placed deliberately, but wind and water carry it in unexpected ways. The perimeter of a prescribed burn is inscribed across a pine savanna, yet the patterns within that boundary are dependent on factors outside of the designer's control. Correspondingly, choreography can be precise in some ways, such as in deciding when an action is taken, but it is not as definitive as building a retaining wall or planting a grove of nursery-grown trees. Choreography requires designers to relinquish some of their control, to be less prescriptive. As Gilles Clément noted, this is quite different from how many other forms of design, such as building architecture, operate:

> 'The architects, the designers, can assume that at the end of the chain, an object ("the project") will appear, whose stable forms will be legible for

> a long time, corresponding to the design submitted. Hard landscaping withstands time. At least a certain time. Not so the garden. It moves with it.
> No sooner installed, it begins to unfold. The designed forms are confirmed. But time works at unraveling the design. It outmaneuvers the knowledgeable speculations, the fine anticipations. It transforms certainties into daily questions and knowledge into a mass of data, juxtaposed according to their academic logic.'
>
> (Clément, 2015, p. 162)

Similarly, in much larger landscapes than a garden, the role of the choreographer is to shape landscape trajectories by way of intentionally introduced, but loosely bound, processes. There are many ways to do this. Fire lines and fence lines act as armatures that processes react to. Dredged sand may be placed in a bay in precise quantities and locations, knowing that waves and currents will move it again. A new course may be excavated for a river, knowing that the river will readily take that excavation as a starting point, but also soon begin to erode banks and build new bars. In actions like these, choreographers leave room for the work of many other agents or (co) designers, who will each act according to their own propensities. The careful intention of the choreographer, in collaboration with the generative chaos of the choreographed, is capable of generating form and instigating landscape changes at a large scale.

In embracing the unruliness of other agents, choreography aligns with the nature of landscape as a medium. Change is fundamental to landscape, both because landscape always unfolds over time (Fulton, 2016; Czerniak, 1997; Howett, 1998; Milligan, 2022) and because the dynamic exchange of matter and energy is crucial to the functioning of ecological systems (Hill, 2005). In recent decades, landscape architects have sought to engage these qualities; however, an overreliance on the methods of practice tied to architecture and building construction has hampered our ability to effectively engage change (Davis and Oles, 2014; Raxworthy, 2019). Embracing choreography can liberate landscape practice from these constraints.

Toward Choreography in Landscape Design

Both architects and landscape architects have occasionally adopted the term choreography. Most often, this has been used to describe the design of human patterns of movement through space (Merriman, 2010; Wasserman, 2012; Gallier, 2015). This version of choreography is relatively closed: The designer aims to sequence movement and space, typically with the intent of producing affective responses in the people who will experience their project. These users are offered little latitude to act as agents in the making of a

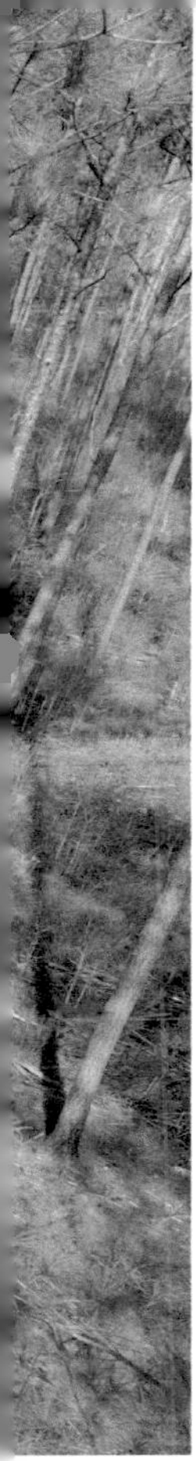

Figure 15.3 | A prescribed burn at the Alabama Meadows Project, designed by David Hill and Emily Knox, generates intricate patterning within the hard boundary of a circular fire line. Photograph by Emily Knox.

landscape; they walk along designed paths and linger in designated rooms, encountering framed views and carefully-organized spatio-atmospheric phenomena. This kind of choreography is worth celebrating. The capacity to design it well is one of the hard-won skills of the experienced landscape architect. But while it is not necessarily incompatible with the openness and multiplicity of agency we have described for our understanding of choreography, it does not facilitate or produce that openness. Consequently, it is perhaps better described as the design of procession than as choreography.

Much of landscape architect Lawrence Halprin's built work—projects such as Seattle's Freeway Park, Portland's Open Space Sequence, or Rochester's Manhattan Square Park—uses choreography in this processional fashion: The orchestrated movement of and interaction between bodies in space (Hirsch, 2014). In later work, documented in 'RSVP Cycles,' he described a relatively novel process for collaboratively developing creative work (Halprin, 2014; Wasserman, 2012). 'RSVP Cycles' offers a dynamic mode of design involving multiple actors. These actors engage through a loosely formulated plan influenced by feedback loops, knowledge sharing, and action (or performance). Halprin described a less prescriptive, ever-evolving way of working, and began to consider the act of design itself as choreography. His changing conception of the design process developed in large part because of Lawrence's long-time collaboration with his creative partner, post-modern dance pioneer Anna Halprin, who was herself a choreographer. But the Halprins' work focused on humans as generative participants in choreography; other beings or processes were not incorporated as agents.

Other work, which often does not use the word choreography itself, lies even closer to the practice we advocate. At the Parc du Sausset, located in a Parisian suburb, landscape architects Michel and Claire Corajoud spent decades cultivating the transformation of fallow farm fields into a rich quilt of fields, meadows, and forests with a large team of collaborators across fields such as forestry and urban agriculture (Di Carlo, 2015). Hundreds of blocks of trees were planted across the park; many remain in place, while other zones were designated as nurseries, growing trees for use in later phases of construction. The blocks that remained in place were treated as in-situ experiments in reforestation, variously thinned, coppiced, or left alone, shaping a diversity of trajectories and spatial characters.

Arquitectura Agronomia deployed a gardening-based approach to choreography in their proposal for Valencia Central Park. The park is animated by a series of maintenance interventions that correspond to seasonal fluxes in phenomena such as plant emergence and insect life. The proposal depends on the designer being in continuous conversation with the landscape, making decisions in response to the continuously shifting planted matrix of the park; for example, paths shift in response to what is in bloom at any given time (Mouritz and Galí-Izard, 2018). At a much different scale, Kienast Vogt designed a courtyard wall at Mühlenbachstrasse, where water pours across the stone of the wall. Over decades, the resulting wetness has fostered an emergent tapestry of soft mosses and ferns that is far more intricate and subtle than it could have been if the plants had been individually specified.

Between the Technical and Poetic

Together with the work of choreographers in other fields (Beiswanger, 1962; Manning, 2013), such projects point toward a field of landscape architecture that employs choreography to expand its expressive potential, work in new territories, substantively claim the mantle of (co)design, and act at scales that the traditional means of landscape practice cannot effectively engage. Landscape architecture has much to gain from broadening the palette of tools and means it uses to make landscapes, and the choreographic tools of fields like forestry, rangeland management, and sediment management should be among those.

Yet landscape architecture is rooted in a set of values and ethics that fundamentally differ from those of other disciplines, who often understand their work as primarily technical and work toward a singular, efficiency-based goal. A typical sediment choreographer chooses when and where sediment is dredged (and where dredged sediment is placed) to maximize the economic efficiency of the work of maintaining navigation channels. A typical rangeland choreographer chooses where to graze to optimize weight gain. A typical forester chooses when to thin in order to generate the most merchantable end product. A landscape architect, on the other hand, is more inclined to consider a multiplicity of intentions for those same flows. We might ask how a more diverse collection of tree species could be choreographed to co-exist in a productive forest, how the herd's movements might meaningfully shift plant assemblages and increase ecological function, or how sediment flows might be guided to produce new publicly-accessible land—to say nothing of the rich atmosphere of the damp woods, the stark expansiveness of the range, or the yellow sublimity of the winter marsh.

A choreographic design practice sacrifices neither the technical nor the poetic but seeks the rich ground where the two overlap. Intentionally working with ecological processes as we have described requires a robust understanding of the behaviors and tendencies of the choreographed materials, whether sand, water, mud, or rangeland grasses and livestock. A skillful choreographer will leverage such technical knowledge in search of the poetic, leaving some degree of form-making up to the will of those materials and agents. A choreographer believes that, given the opportunity, those agents have the potential to be generative, to create intricate new landscape forms and patterns through their own dynamic tendencies. A choreographer does not seek a singular future state but understands that their work is never final or complete and that revisions can and should always be made. Choreography prefers landscape care to landscape control; it allows a designer to be in long-term conversation with a landscape and return again and again to repeatedly mend and amend. By working at the scale of the herd, the burn, or the flow of water and sediment, choreography allows designers to engage with much larger landscapes than typical landscape construction. It invigorates our practice by opening it up to an animated world—one where we can see forests, cattle, mushrooms, gravel, and forbs not merely as matter to enforce our wills upon, but as collaborators in design.

Notes

1 | In this chapter, we follow Marc Steen in adopting Elizabeth B.-N. Sanders and Pieter Jan Stappers' definition of co-design as 'collective creativity as it is applied across the whole span of a design process' (Steen, 2013; Sanders and Stappers, 2008).

2 | Teresa Galí-Izard's *The Same Landscapes* provides a template for this exercise since she examines landscape-making acts such as quarrying, hunting, composting, and pruning on level ground with the work of landscape architects sketching the contours of a landscape architecture grounded in broadly-gathered techniques and a deep understanding of landscapes as a medium. Galí-Izard specifically recommended considering landscape projects in terms of choreography, suggesting that their maintenance operations be thought of 'as a type of choreography for dancers moving about the garden' (Galí-Izard and Simon, 2005).

3 | Or, at least, very rarely is it like those worlds. We discuss some counterexamples of choreographic landscape architecture at the end of this essay.

4 | This ideology of control has also affected the work of landscape-makers in other domains outside of landscape architecture. Landscape-makers working with sediment and on rangelands have an all-too similar set of tendencies, constructing hard infrastructure like seawalls and levees, enforcing uniformly hard edges that effectively disconnect landscape from process.

5 | More recently, Billy Fleming makes a similar point in 'Design and the Green New Deal': 'we [landscape architects] may yearn to impart systems-level change, but we are working on discrete sites, with incrementalist tools, within structures that produce injustice' (Fleming, 2019).

References

Beiswanger, G. (1962) 'Chance and design in choreography.' *The Journal of Aesthetics and Art Criticism*, 21 (1), pp. 13–17.

Bélanger, P. (2015) 'Is landscape infrastructure?' In: Doherty, G. and Waldheim, C. eds. *Is Landscape…?* New York, Routledge, pp. 190–227.

Briske, D. (2017) *Rangeland Systems: Processes, Management and Challenges*. Germany, Springer International Publishing, 2017.

Cantrell, B., Zhang, Z. and Liu, X. (2021) 'Artificial intelligence and machine learning in landscape architecture.' In: As, I. and Basu, P. eds. *The Routledge Companion to Artificial Intelligence in Architecture*. London, Routledge, pp. 232–247.

Clément, G. (2015) 'The Wisdom of the Gardener.' In: *"The Planetary Garden" and Other Writings*. Philadelphia, University of Pennsylvania Press, pp. 1–64.

Cook, R.E. (2000) 'Do landscapes learn? Ecology's new paradigm and design in landscape architecture.' In: Conan, M. ed. *Environmentalism in Landscape Architecture*. Washington, Dumbarton Oaks, pp. 115–132.

Coomans, T., Cattoor, B. and De Jonge, K. (2019) *Mapping Landscapes in Transformation Multidisciplinary Methods for Historical Analysis*. Leuven, Leuven University Press.

Corner, James (1997). 'Ecology and landscape as agents of creativity.' In: Steiner, F. and Thomson, G. eds. *Ecological Design and Planning*. New York, Wiley, pp. 80–108.

Czerniak, J. (1997) 'Challenging the pictorial: recent landscape practice.' *Assemblage*, 34, pp. 110–120.

Davis, B. and Oles, T. (2014) 'From Architecture to Landscape.' *Places Journal*.

Dee, C. (2010) 'Form, utility, and the aesthetics of thrift in design education.' *Landscape Journal*, 29 (1), pp. 21–35.

De la Peña, D., Allen, D.J., Hester, R.T., Hou, J., Lawson, L.L. and McNally, M.J. eds. (2017) *Design as Democracy: Techniques for Collective Creativity*. Washington, DC, Island Press.

Di Carlo, F. (2015) 'Michel Corajoud and Parc Départemental du Sausset.' *Journal of Landscape Architecture*, 10 (3), pp. 68–77.

Fleming, B. (2019) 'Design and the green new deal.' *Places Journal*. Available at: <https://placesjournal.org/article/design-and-the-green-new-deal/> [Accessed 10 November 2022].

Fulton, G. (2016) 'How the garden moves.' *Landscape Architecture Magazine* 106 (6), pp. 152–154.

Funtowicz, S.O. and Ravetz, J.R. (1994) 'Uncertainty, complexity and post–normal science.' *Environmental Toxicology and Chemistry: An International Journal*, 13 (12), pp. 1881–1885.

Halprin, L. (2014) 'The RSVP cycles: creative processes in the Human environment.' *Choreographic Practices*, 5 (1), pp. 39–47.

Hays, S.P. (1999) *Conservation and the Gospel of Efficiency: The Progressive Conservation Movement, 1890–1920*. Pittsburgh, University of Pittsburgh Press.

Hester Jr., R.T. (2010) *Design for Ecological Democracy*. Cambridge, MIT Press.

Hill, K. (2005) 'Shifting sites.' In: Burns, C. and Kahn, A. eds. *Site Matters*. London, Routledge, pp. 161–163.

Hirsch, A.B. (2014) *City Choreographer: Lawrence Halprin in Urban Renewal America*. Minneapolis, University of Minnesota Press.

Holmes, R., Milligan, B. and Wirth, G. (forthcoming) *Silt Sand Slurry: Sediment, Infrastructure, and the Worlds that We are Making*. San Francisco, ORO Applied Research + Design.

Holmes, R. (2020) 'The problem with solutions.' *Places Journal*. Available at: <https://placesjournal.org/article/the-problem-with-solutions/> [Accessed 10 November 2022].

Howett, C. (1998) 'Ecological values in twentieth-century landscape design: A history and hermeneutics.' *Landscape Journal*, 17, pp. 80–98.

Galí-Izard, T. and Simon, J. (2005) *The Same Landscapes: Ideas and Interpretations*. Barcelona, Gustavo Gili.

Gallier, E. (2015) 'Remembrance Park 14–18: The landscape as a score for the choreography of memory.' *Journal of Landscape Architecture*, 10 (1), pp. 38–47.

Geffel, M. (2021) 'Landscape design through maintenance: field case studies in parametric mowing.' *Landscape Journal*, 39 (2), pp. 1–16.

Girot, C. (2021) *Choreografien der Landschaft*. Available at: <https://choreografienderlandschaft.cargo.site/Choreography> [Accessed 20 January 2023].

Knox, E. (2022) 'Choreographing desire.' *Wonderground Journal*, 3, pp. 107–115.

Lokman, K. (2017) 'Cyborg landscapes: Choreographing resilient interactions between infrastructure, ecology, and society.' *Journal of Landscape Architecture*, 12 (1), pp. 60–73.

Manning, E. (2013) *Always More Than One: Individuation's Dance*. Durham, Duke University Press.

Mathur, A. and Da Cunha, D. (2009) *SOAK: Mumbai in an Estuary*. New Delhi, Rupa & Company.

Merriman, P. (2010) 'Architecture/dance: Choreographing and inhabiting spaces with Anna and Lawrence Halprin.' *Cultural Geographies*, 17 (4), pp. 427–449.

Milligan, B. (2022) 'Accelerated and decelerated landscapes.' *Places Journal*. Available at: <https://placesjournal.org/article/accelerated-and-decelerated-landscapes/> [Accessed 10 November 2022].

Milligan, B. (2015) 'Landscape migration.' *Places Journal*. Available at: <https://placesjournal.org/article/landscape-migration/> [Accessed 10 November 2022].

Mouritz, L. and Gali-Izard, T. (2018) 'Teresa Gali-Izard: The language of landscape.' *Landscape Architecture Australia*, 158, pp. 46–51.

Raxworthy, J. (2019) *Overgrown: Practices between Landscape Architecture and Gardening*. Cambridge, MIT Press.

Rosati, J.D., Carlson, B.D., Davis, J.E. and Smith, T.D. (2004) *The Corps of Engineers National Regional Sediment Management Demonstration Program*. Vicksburg, MS, Engineer Research and Development Center, Coastal and Hydraulics Lab.

Sanders, E.B.N. and Stappers, P.J. (2008) 'Co-creation and the new landscapes of design.' *Co-design*, 4 (1), pp. 5–18.

Sayre, N.F. (2017) *The Politics of Scale*. Chicago, University of Chicago Press.

Schröder, T. (2001) 'Put not your trust in naive gardens!' In: Schröder, T., *Changes in Scenery: Contemporary Landscape Architecture in Europe*. Basel, Birkhäuser, pp. 78–89.

Shallat, T. (2010) *Structures in the Stream: Water, Science, and the Rise of the US Army Corps of Engineers*. Austin, University of Texas Press.

Solan, V. (2017) 'Living lessons.' *Landscape Architecture Magazine*, 107 (3), pp. 24–26.

Steen, M. (2013) 'Co-design as a process of joint inquiry and imagination.' *Design Issues*, 29 (2), pp. 16–28.

Wasserman, J. (2012) 'A world in motion: the creative synergy of Lawrence and Anna Halprin, *Landscape Journal*, 31 (1-2), pp. 33–52.

Wu, J. and Loucks, O.L. (1995) 'From balance of nature to hierarchical patch dynamics: a paradigm shift in ecology.' *The Quarterly Review of Biology*, 70 (4), pp. 439–466.

Index

For Product Safety Concerns and Information please contact our EU representative GPSR@taylorandfrancis.com Taylor & Francis Verlag GmbH, Kaufingerstraße 24, 80331 München, Germany

Batch number: 10733620

Printed by Printforce, the Netherlands